The mathematical theory
of quantitative genetics

# THE MATHEMATICAL THEORY OF QUANTITATIVE GENETICS

M. G. BULMER

Lecturer in Biomathematics,
University of Oxford

CLARENDON PRESS · OXFORD
1980

*Oxford University Press, Walton Street, Oxford* OX2 6 DP

OXFORD  LONDON  GLASGOW
NEW YORK  TORONTO  MELBOURNE  WELLINGTON
KUALA LUMPUR  SINGAPORE  JAKARTA  HONG KONG  TOKYO
DELHI  BOMBAY  CALCUTTA  MADRAS  KARACHI
IBADAN  NAIROBI  DAR ES SALAAM  CAPE TOWN


Published in the United States by Oxford University Press, New York

**British Library Cataloguing in Publication Data**

Bulmer, Michael George
  The mathematical theory of quantitative
  genetics.
  1. Genetics—Mathematics
  I. Title
575.1 '01 '8     QH438.4.M33     80-40338

ISBN 0-19-857530-0


Printed and bound by
Spottiswoode Ballantyne Ltd,
Colchester and London.

# PREFACE

Classical genetics considers the inheritance of discrete characters, such as the presence or absence of horns in cattle, for which individuals can be classified unambiguously into a small number of distinct types. In the simplest case the character is determined by a single gene and is unaffected by environmental variability. By contrast, many characters of agricultural and biological importance, such as milk yield in cows or wing span in birds, present a continuous range of variability, and they may be influenced by a large number of genes of individually small effect as well as by different environmental factors. Quantitative genetics is the study of the inheritance of such continuous, quantitative characters.

Quantitative genetics is based on the assumption that continuous characters are determined by genes which behave in the same way as the genes of major effect which control discrete characters. The existence and properties of these genes cannot be demonstrated directly by classical genetic methods since distinct, non-overlapping classes corresponding to different genotypes do not exist. They must instead be inferred from statistical properties of the character, such as the change in the mean and variance in successive generations following a cross between two inbred lines, correlations between relatives in an outbred population, and the effects of inbreeding and selection on the mean of the character.

The purpose of this book is to develop a logical account of the underlying theory of quantitative genetics. This theory provides the bridge between the observable statistical properties of the character and the genetic factors which, together with environmental factors, are postulated to determine the expression of the character. The structure of the book is as follows: a brief account of classical, Mendelian genetics and of the pioneer experiments in quantitative genetics (Chapter 1); statistical description of the role of genotype and environment and their interaction in determining the observed phenotypic value of a character (Chapter 2); derivation of the statistical model for decomposing the genotypic value into components representing different types of gene action (Chapters 3 and 4); estimation of the components of variance and similar quantities generated by this model from biometric analysis of different types of experiments and observations in inbreeding and outbreeding populations (Chapters 5–7); the role of the normal distribution in quantitative genetics (Chapter 8); the effect of selection on a quantitative character, and the applications of selection theory in animal and plant breeding and in evolutionary theory (Chapters 9–11); stochastic effects due to finite population size (Chapter 12).

The emphasis is on discussion of the underlying principles, illustrated by simple examples. It is hoped that a clear understanding of basic principles in simple cases will enable the reader to extend the results to deal with the many complexities which arise in practice, though only a few of them are considered here. The reader is assumed to have some knowledge of statistical methods and theory, particularly in the area of regression and analysis of variance on which quantitative genetics rests so heavily. I have, however, tried to make the book intelligible to the biologist with some statistical experience as well as to the statistician; some of the more difficult or more specialized material has been starred, and may be omitted at first reading without loss of continuity. Problems for solution are given at the end of each chapter.

I have had valuable discussions with the following while writing this book: Mike Arnold, Peter Avery, Bill Hill, Rod Kempton, Russ Lande, Tom Nagylaki, Alan Robertson, Charles Smith, John Snape, Robin Thompson, and Tony Wright. I am grateful to Peter Avery and Oliver Mayo for constructive criticism of the manuscript.

*Oxford*  MGB
June 1979

# CONTENTS

# x    Contents

# 1 The biological background

*Quantitative genetics is the study of the inheritance of quantitative characters. It is based on the assumption that such characters are determined by genes which behave in the same way as the genes of major effect which control discrete characters. This introductory chapter is intended to give a brief review of the main concepts of classical, Mendelian genetics and of their application to quantitative characters. For further details the reader is referred to a standard textbook, such as Strickberger (1976) or Whitehouse (1973).*

### Mendel's experiments

The science of genetics began with Mendel's experiments on garden peas. As an example, we shall consider his experiments on flower colour. He had obtained from seedsmen two varieties of pea, one with purple flowers and the other with white flowers, and he verified that this character difference remained constant over several generations; one would expect different varieties to breed true since the pea is self-fertilizing. Mendel then crossed these two varieties by dusting pollen from one variety onto the stigma of the other; whichever way round this cross was done all the resulting plants in the next ($F_1$) generation had purple flowers. However, when these $F_1$ plants were allowed to self-fertilize, only three-quarters of the plants in the next ($F_2$) generation had purple flowers, and the remaining quarter had white flowers; the actual numbers of $F_2$ plants were 705 with purple and 224 with white flowers, giving a proportion of 0.759 with purple flowers.

Mendel's explanation of these results is shown diagrammatically in Fig. 1.1.

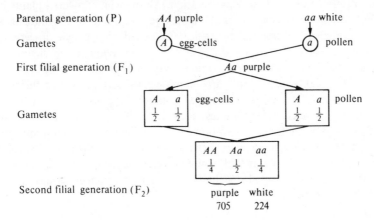

Fig. 1.1 Interpretation of Mendel's experiment on flower colour in garden peas.

## 2 The biological background

The following assumptions are made: (1) flower colour is determined by a factor (today called a gene) which can exist in two forms (alleles), $A$ and $a$, determining purple and white flowers respectively; (2) each plant contains two of these genes, one derived from each of its parents, so that it can have one of the three possible genotypes, $AA$, $Aa$, or $aa$; (3) the gametes (egg-cells and pollen, or ova and sperm in animals) contain only one gene chosen at random from the two genes of the parent plant; (4) the egg-cells and pollen unite at random, independently of the genes they carry, to produce the plants of the next generation. The essential feature of this theory is the separation or segregation of allelic genes when the gametes are formed.

In a self-fertilizing species we expect all plants to be homozygous (either $AA$ or $aa$), as will become clear shortly. Thus when the two true-breeding varieties were crossed at the beginning of the experiment the mating was between two homozygotes, $AA \times aa$; all the resulting plants must have had the heterozygous genotype, $Aa$, since they received $A$ from the first and $a$ from the second parent. In order to account for the fact that all these plants had purple flowers like the first parent, we must suppose that both $AA$ and $Aa$ plants have purple flowers and only $aa$ plants have white flowers. Purple flower colour is said to be dominant and white recessive. A plausible explanation might be that the $A$ gene leads to the production of a purple pigment, enough of which is produced to make the flowers purple whether the gene is present in single or in double dose. It is useful to draw a distinction between the genotype of a plant ($AA$, $Aa$, or $aa$) and its phenotype (purple or white flowers).

When the heterozygous $F_1$ plants were allowed to self-fertilize the mating was of the type $Aa \times Aa$. In this case egg-cells containing $A$ and $a$ will be produced with equal probabilities, and likewise for the pollen. Hence the probabilities of the genotypes $AA$, $Aa$, and $aa$ in the next generation will be $\frac{1}{4}, \frac{1}{2}$, and $\frac{1}{4}$ respectively, bearing in mind that the heterozygote $Aa$ can arise in two ways ($A$ egg-cell and $a$ pollen, or vice versa). Thus three-quarters of them will have purple flowers and one-quarter white flowers, as observed.

Mendel confirmed this theory in two ways. Firstly, he allowed the $F_2$ plants to self-fertilize to produce an $F_3$ generation. All the white-flowered plants bred true, but among the purple-flowered plants, one-third bred true, producing only purple-flowered offspring, and two thirds produced both purple-flowered and white-flowered offspring in a ratio of 3:1. This behaviour is predicted, since white-flowered plants are always homozygous $aa$, whereas among the purple-flowered $F_2$ plants, one-third are homozygous $AA$ and two-thirds are heterozygous $Aa$. (It should be observed that homozygotes breed true on self-fertilization, whereas heterozygotes produce half heterozygotes and one-quarter of each of the two types of homozygotes. Thus the overall frequency of heterozygotes is halved in each generation of self-fertilization.) Secondly, Mendel crossed the $F_1$ plants with white-flowered plants in a backcross experiment, and found as expected that half the progeny had purple flowers and the other half white flowers.

Mendel examined seven contrasting pairs of characters (such as round v. wrinkled seeds and tall v. short stem) in the same way, and he obtained similar results in each case, with one character being completely dominant to the other. He also investigated what happened when plants differing in two contrasting characters were crossed. Having established that round seeds were dominant to wrinkled seeds and yellow seed colour to green seed colour, he crossed a true-breeding variety having round, yellow seeds with a variety having wrinkled, green seeds; the resulting $F_1$ seeds were all round and yellow as expected. (Note that both these seed characters are determined by the genotype of the seed, not by that of the mother plant.) The plants grown from these seeds were then either allowed to self-fertilize to produce $F_2$ seeds or crossed to the variety with wrinkled, green seeds in a backcross. The results of these two experiments are shown in Table 1.1.

TABLE 1.1

*Mendel's data on the joint segregation of seed shape and colour in the garden pea*

|  |  | Yellow | Green | Total |
|---|---|---|---|---|
| $F_2$ | Round | 315 | 108 | 423 |
|  | Wrinkled | 101 | 32 | 133 |
|  | Total | 416 | 140 | 556 |
|  |  | Yellow | Green | Total |
| Backcross | Round | 55 | 51 | 106 |
|  | Wrinkled | 49 | 52 | 101 |
|  | Total | 104 | 103 | 207 |

The marginal totals are in the ratio of $3:1$ in the $F_2$ experiment and $1:1$ in the backcross as expected. Furthermore, it is clear that the two characters are behaving independently of each other in the sense that there is no tendency for round seeds to be yellow and wrinkled seeds green, or vice versa. It can be concluded that the genes for these two characters segregate independently of each other when the gametes are formed. Write $A$ and $a$ for the genes determining round and wrinkled seeds, and $B$ and $b$ for the genes determining yellow and green seed respectively. The $F_1$ seeds have the double heterozygous genotype $AaBb$, and the plants grown from them will produce four gametic types, $AB$, $Ab$, $aB$, and $ab$. The results in Table 1.1 can be explained by supposing that a gamete receives one of the two seed-shape genes and one of the two seed-colour genes at random and independently of each other; in consequence the four gametic types should occur with the same probability of $\frac{1}{4}$. However, it will be shown in the next section that there are exceptions to the rule of the independent segregation of genes for different characters.

Mendel's paper was published in 1866 but its importance went unrecognized until it was simultaneously rediscovered by three biologists in 1900. It was quickly shown that Mendel's theory could be used to explain the inheritance of many characters in both plants and animals, but that some extension of the theory was required. Thus it soon became clear that complete dominance, though common, was not universal. For example, roan cattle are heterozygous for a pair of alleles which in homozygous form produce either red or white coat-colour. Thus red cattle mated together produce red calves, white cattle mated together produce white calves, red bulls mated to white cows (or vice versa) produce roan calves, and roan cattle mated together produce red, roan, and white calves in an average ratio of $1:2:1$.

Another complicating factor was the discovery of interaction (or epistasis) between genes. Bateson, Saunders, and Punnett (1905) crossed two white-flowered varieties of the sweet pea, and found to their surprise that all the $F_1$ progeny had purple flowers. On self-fertilization three flower colours appeared in $F_2$, purple, red, and white, in the approximate ratios of $27:9:28$. These ratios together with further breeding data showed that flower colour was determined by three independently segregating genes $A$, $B$, and $C$ with recessive alleles $a$, $b$, and $c$, respectively. Colour is only produced if both $A$ and $B$ are present (in single or double dose); the colour produced in non-white flowers is purple or red depending on the presence or absence of $C$. The experimental results can be explained by supposing that the original cross was of the type $AAbbCC \times aaBBcc$. A possible biochemical explanation of how these genes act is shown below; it is supposed that each reaction can only occur in the presence of the appropriate gene.

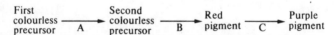

First colourless precursor → (A) Second colourless precursor → (B) Red pigment → (C) Purple pigment

Fig. 1.2. Possible biochemical mechanism for the inheritance of flower colour in the sweet pea.

**Chromosomes**

The genes were originally hypothetical constructs invented to account for the results of breeding experiments. It was soon realized that there are physical objects, the chromosomes in the nuclei of cells, which behave in exactly the way in which genes are postulated to behave; it therefore became natural to suggest that the genes are carried on the chromosomes.

Briefly, every somatic cell nucleus contains a number of chromosomes which is characteristic of the species of animal or plant; there are 46 chromosomes in every human cell, 40 in the mouse, 12 in the housefly, 14 in the garden pea, 20 in maize, and so on. These chromosomes can be arranged in pairs, the members of each pair (homologous chromosomes) being physically identical, but distinguishable in shape, size, or some other characteristic from the members of any other pair. (The sex chromosomes are the exception to this rule; they will be discussed later.) In ordinary cell-division each chromosome is replicated exactly, so that the daughter cells contain the same number of chromosomes as the parent cell. However, the gametes are produced by a special type of cell-division called meiosis in which each cell divides twice to produce four daughter cells, but the chromosomes are replicated once. In consequence, the chromosome number in the gametes is halved, only one chromosome from each homologous pair being represented in each gamete; such cells are called haploid. When two gametes unite to form the zygote the original (diploid) complement of chromosomes is restored. These cytological facts provide the physical basis for Mendelian inheritance (cf. Fig. 1.1).

The theory that the genes are carried on the chromosomes explained the phenomenon of linkage which had puzzled the early geneticists. The independent segregation of seed shape and colour in Table 1.1 can be explained by supposing that the genes controlling these factors are on different (non-homologous) chromosomes, which segregate independently at meiosis. After the rediscovery of Mendelism in 1900, it was soon found that most characters segregate independently of each other in this way, but that there are some exceptions, which can be explained by supposing that the corresponding genes are at different positions (loci ) on the same chromosome; such genes are said to be linked.

As an example Table 1.2 shows the results of a backcross experiment on two seed characters in maize, coloured v. colourless and starchy v. waxy seeds. A pure-breeding variety of maize with coloured, starchy seeds was crossed to a variety with colourless, waxy seeds; all the resulting $F_1$ seeds were coloured and starchy, showing the dominance of these two characters. The plants grown from these seeds were then crossed back to the variety with colourless, waxy seeds with the results shown in Table 1.2. Both the marginal totals show good 1:1 segregations, but the two factors are clearly not behaving independently of each other.

TABLE 1.2

*The joint segregation of two seed characters in a backcross experiment in maize (Bregger 1918; quoted by Whitehouse 1973)*

|  | Starchy | Waxy | Total |
| --- | --- | --- | --- |
| Coloured | 147 | 65 | 212 |
| Colourless | 58 | 133 | 191 |
| Total | 205 | 198 | 403 |

We denote the allelic genes for colour and lack of colour by $A$ and $a$, and the genes for starchiness and waxiness by $B$ and $b$ respectively, and we suppose that these genes are at different loci on the same chromosome. The genotype of the $F_1$ seeds may be represented as $AB/ab$, indicating that $A$ and $B$ are on one chromosome, inherited from the first parent, while $a$ and $b$ are on the homologous chromosome inherited from the second parent. During meiosis homologous chromosomes come together and lie side by side; breaks then occur, at the same randomly located places in both chromosomes, and the broken pieces rejoin with their opposite partners. This process is called crossing-over, and its effect is that the chromosomes in the gametes are mosaics of the homologous maternal and paternal chromosomes in the parent organism (Fig. 1.3). An even number of cross-overs between two loci has no net effect, but an odd number of cross-overs will change $AB/ab$ into $Ab/aB$ (or vice versa). This is called recombination, and the probability that it will occur is the recombination fraction, denoted by $r$. Thus the double heterozygote $AB/ab$ will produce four types of gametes, the two recombinant types $Ab$ and $aB$ with probability $\frac{1}{2}r$ each and the two non-recombinant types $AB$ and $ab$ with probability $\frac{1}{2}(1 - r)$ each. In a backcross to the double recessive these four gametic types can be recognized directly; in Table 1.2 the two recombinant types are coloured waxy and colourless starchy, and the two non-recombinant types are the other two, so that the recombination fraction can be estimated as $r = (58 + 65)/403 = 0.31$.

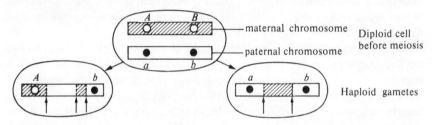

Fig. 1.3. Simplified diagram of the effects of crossing-over; arrows mark the points at which cross-overs have occurred.

In organisms which have been extensively investigated, such as the fruitfly *Drosophila melanogaster* and maize, it has been found that the genes fall into linkage groups, the genes in each group being linked to each other but not to genes in any other group. The number of linkage groups is equal to the haploid number of chromosomes, four in *Drosophila* and ten in maize; this fact provides strong support for the theory that the genes lie on the chromosomes. Furthermore, one would expect the recombination fraction between two linked loci to depend on the distance between them on the chromosome; loci close together will seldom recombine, whereas loci far apart on the same chromosome will behave almost as if they are unlinked, with a recombination fraction nearly $\frac{1}{2}$. Linkage data have been successfully used in several species to construct a consistent linear map of the chromosomes, in which the order of the genes and their approximate distances apart are determined.

We must now consider the sex chromosomes which have so far been ignored. In species which have separate sexes (most animals and a few plants) the sex of an individual is determined by a pair of sex chromosomes, denoted X and Y. In most species the female has two X chromosomes, whereas the male has one X and one Y chromosome; the sex chromosomes in males are the exception to the rule that the chromosomes occur in homologous pairs. (The other chromosomes, which occur in homologous pairs in both sexes, are called autosomes.) All the ova produced by females contain a single X chromosome, but the male produces X-bearing and Y-bearing sperm in equal numbers; an X-bearing sperm when it fertilizes an ovum will produce an XX female zygote, whereas a Y-bearing sperm will produce an XY male zygote. (This is the typical genetic mechanism for determining sex, though there are several variants on it; for example, the situation is reversed in birds, butterflies, and moths, males being XX and females XY.)

The Y chromosome is genetically almost inert, apart from its role in determining sex, but the X chromosome has genes for other characters on it; such genes are said to be sex-linked. A well-known human sex-linked character is colour-blindness, whose inheritance can be explained on the following assumptions: (1) there is a pair of allelic genes, $C$ and $c$, on the X chromosome determining colour vision; (2) in females normal colour vision is dominant to colour blindness, so that both $CC$ and $Cc$ women have normal vision and only $cc$ women are colour-blind; (3) there is no analogue for this gene on the Y chromosome, so that men have only one gene for colour vision, $C$ men being normal and $c$ men colour-blind. This model accounts for the main features in the inheritance of colour blindness: (1) it is commoner in men than women because only one $c$ gene is required to produce colour blindness in men whereas two $c$ genes are required in women; (2) if a normal man marries a colour-blind woman, all their sons and none of their daughters are colour-blind; (3) if a colour-blind man marries a normal woman, she will usually be homozygous $CC$; in this case all their children will be normal, but their daughters will be heterozygous carriers, $Cc$, and if they marry a normal man, then half their sons (but none of their daughters) will be colour-blind.

### The nature of the gene

The classical analogy is to liken the chromosome to a string of beads, each bead representing a different gene. This model incorporates the two main features of classical genetics, the particulate nature of genes and their linear arrangement on chromosomes. Though it must be modified in detail, the string of beads still provides a valuable physical model of how genes behave. However, modern work in molecular biology has greatly increased our understanding of what genes are and how they work at the biochemical level.

The main chemical constituent of nuclei is a complex substance called nucleoprotein consisting of DNA (deoxyribonucleic acid) associated with protein. It was at first assumed that the genetic information was carried in the protein fraction, but it became clear in the 1940s that it was in fact carried in the nucleic acid fraction of nucleoprotein; two lines of evidence were the transfer of genetic information by purified DNA in bacteria (bacterial transformation) and the fact that the DNA content of cell nuclei was constant in somatic cells from different tissues and was exactly twice the content of the gametes. The breakthrough in understanding how DNA carries genetic information came with the elucidation of its molecular structure by Watson and Crick in 1953.

DNA is a very large molecule built of components called nucleotides; a nucleotide consists of one of four possible nitrogenous bases (adenine, guanine, thymine, or cytosine, usually abbreviated as A, G, T, or C) to which is attached a sugar (deoxyribose) and a phosphate group. Watson and Crick proposed that chromosomal DNA is composed of two strands of linked nucleotides, an A in one strand always being paired with a T in the other, and a G with a C. The base composition might therefore look as follows:

$$\begin{array}{l} \text{A A G T C G G T C} \ldots \\ \text{T T C A G C C A G} \ldots \end{array} \qquad (1.1)$$

This theory was based on building chemical models to interpret X-ray studies of crystalline DNA, together with the observation that although the proportions of the four bases varied in DNA from different species, the A content and the T content were always the same, as were the G and C contents.

The Watson–Crick model of the structure of DNA has two important biological implications. First, it immediately suggests a mechanism for the self-replication of DNA; if the two strands are split open, each will attract to itself a new complementary strand because of the mutual pairwise affinities of the bases. Secondly, it suggests that the genetic information in DNA is carried by the sequence of bases. How is this information translated into the observed phenotypic effects of genes?

It has been realized for a long time that genes exert their primary effect through the production of proteins, in particular of enzymes which catalyse the many chemical reactions taking place in the body. This has been summarized in the phrase 'one gene–one enzyme', which expresses the hypothesis that each gene is responsible for the production of a specific enzyme (though it should be extended to 'one gene–one protein' to include genes which are responsible for proteins which are not enzymes). For example, Garrod (1909) showed that the rare human disease alcaptonuria ('black urine disease') was due to a recessive gene and that it was caused by an inability to break down homogentistic acid so that this substance was excreted in the urine. He suggested that the fundamental defect was the absence of a specific enzyme for metabolizing homogentistic acid; this enzyme has now been isolated and its absence in alcaptonurics confirmed (see Harris 1975). We may suppose that the normal gene produces functional enzyme, and that the abnormal or mutant allele produces an abnormal form of the enzyme which is functionless; the heterozygote apparently has enough of the normal enzyme to function normally. The mechanism of the inheritance of flower colour in the sweet pea in Fig. 1.2 can be interpreted in the same way, the dominant gene at each locus producing an enzyme which catalyses a specific reaction.

A protein is a large molecule built up of components called amino-acids linked together in a chain; twenty different amino-acids occur naturally. To understand how genes produce their effects, we must therefore discover how the linear sequence of nucleotide bases in DNA is translated into the linear sequence of amino-acids in the corresponding protein; in brief, what is the genetic code for translating the DNA message with its four-letter alphabet into the protein message with its twenty-letter alphabet? The answer turned out to be very simple in principle. The code is a triplet code, each triplet of three nucleotides coding for a specific amino-acid. For example, AAA and AAG both code for phenylalanine, GTT and GTC both code for glutamine, and any of the six triplets, AGA, AGG, AGT, AGC, TCA, TCG code for serine. Only one of the two strands of DNA is translated, and this strand is read consecutively in non-overlapping triplets. Thus the top strand in (1.1) would be ·translated into a protein containing the amino-acids phenylalanine–serine– glutamine . . .. There are also signals for starting and stopping the message. A typical protein contains about 150 amino-acids, so that a typical gene may be thought of as a sequence of about 450 nucleotides, beginning with a start signal and ending with a stop signal, together with the complementary chain which is not translated. The gene leads to the production of a particular protein, having a specific, determined sequence of amino-acids.

The above theory gives a new insight into the nature of mutations which produce new alleles and hence generate genetic variability. Occasional mistakes may occur in the replication of DNA. Such mistakes will usually involve single bases, so that for example AAG might become AGG in the twenty-third triplet of a particular gene; the twenty-third amino-acid of the corresponding protein would be changed from phenylalanine to serine. It seems likely that most mutations are of this kind. A well-known example is sickle-cell haemoglobin which differs from normal haemoglobin only in having valine substituted for glutamic acid in the sixth position (out of 146 amino-acids) in the beta chain. (Sickle-cell haemoglobin is less soluble than normal haemoglobin, which leads to severe anaemia in individuals who are homozygous for the sickle-cell gene; heterozygotes have a mixture of the two types of haemoglobin, which seems to give them some protection against malaria, and hence maintains the gene in the population by heterozygous advantage in malarial parts of the world.) It is clear that a very large number of different mutations is possible at each locus since a mistake may occur at any one of the nucleotide sites. It is an open question how many of them are likely to be severely deleterious through causing loss of biological activity and how many may be selectively neutral or even mildly beneficial.

Finally, we will consider dosage compensation at sex-linked loci. At autosomal loci the amount of gene product is proportional to the number of active genes; for example, heterozygotes for the alcaptonuria gene have about half the enzyme activity for metabolizing homogentistic acid as normal homozygotes. If this rule held for sex-linked loci then females would produce twice as much gene product at all loci on the X chromosome as males, which would lead to metabolic imbalance between the sexes. In fact, males and females have the same level of activity for enzymes controlled by sex-linked loci. This phenomenon is called dosage compensation; it extends to all X-linked genes not involved in sex determination.

Different ways of achieving dosage compensation have evolved in different groups of animals. In placental mammals it is brought about by the inactivation of one of the two X chromosomes in each female cell at any early stage in the development of the embryo. It is a matter of chance whether the paternal or the maternal X chromosome is inactivated, so that female mammals are a mosaic, some cells having an active paternal and others an active maternal X chromosome. This is illustrated by the tortoise-shell cat which is phenotypically a mosaic of black and ginger fur (or tabby and ginger fur), and which is genetically a heterozygote for a sex-linked gene producing ginger fur. In kangaroos and other marsupials there is a similar mechanism but it is always the paternal X chromosome which is inactivated. In *Drosophila* it seems that dosage compensation has been achieved in a different way by the evolution of separate modifiers for different genes. These modifiers depress the activity of sex-linked genes in females so that each gene in the female produces half as much gene product as the single gene in the male.

### Quantitative characters

*The multiple factor hypothesis*

Mendel's success in obtaining simple, clear-cut results derived in part from his careful choice of simple, discrete characters to investigate. When his experimental methods were applied to quantitative characters, results were often obtained which required a more complicated hypothesis to explain and more sophisticated methods to analyse. A good example is provided by an experiment of East (1916) on two varieties of an ornamental species of tobacco with different flower lengths. Table 1.3 shows the frequency distribution of flower length in the two parental variations, in the $F_1$ cross between then, and in the $F_2$ generation obtained by allowing $F_1$ plants to self-fertilize. Tobacco is normally self-fertilizing so that the experiment is comparable with Mendel's experiments. (Results for three different years have been pooled since there is no evidence of any difference between them.)

Since tobacco is self-fertilizing the two parents may be assumed to be homozygous at all loci; the $F_1$ plants will all be heterozygous at any loci at which the two parents differ. Thus the variability in the parents and in $F_1$ must be entirely of environmental origin, but the great increase in variability in $F_2$, which is typical of such experiments, can be attributed to the segregation of Mendelian genes. If only a single gene were involved, then the $F_2$ distribution

TABLE 1.3

*Frequency distributions of flower length in a cross between two varieties of tobacco (East 1916)*

| Flower length (mm) (class centre) | $P_1$ | $P_2$ | $F_1$ | $F_2$ |
|---|---|---|---|---|
| 34 | | 1 | | |
| 37 | | 21 | | |
| 40 | | 140 | | |
| 43 | | 49 | | |
| 46 | | | | |
| 49 | | | | |
| 52 | | | | 3 |
| 55 | | | 4 | 9 |
| 58 | | | 10 | 18 |
| 61 | | | 41 | 47 |
| 64 | | | 75 | 55 |
| 67 | | | 40 | 93 |
| 70 | | | 3 | 75 |
| 73 | | | | 60 |
| 76 | | | | 43 |
| 79 | | | | 25 |
| 82 | | | | 7 |
| 85 | | | | 8 |
| 88 | 13 | | | 1 |
| 91 | 45 | | | |
| 94 | 91 | | | |
| 97 | 19 | | | |
| 100 | 1 | | | |
| Total number | 169 | 211 | 173 | 444 |
| Mean | 93.1 | 40.4 | 63.5 | 68.8 |
| Variance | 4.9 | 2.3 | 7.9 | 41.6 |

should be a mixture of the distributions of $P_1$, $F_1$, and $P_2$ in proportions $1:2:1$; thus the $F_2$ distribution should consist of three non-overlapping distributions which is clearly not the case. East and others suggested that the facts could be explained by supposing that the parents differed in several genes controlling the character; this became known as the multiple factor hypothesis.

Suppose that the two parents differ at $n$ loci which affect flower length. Assume for simplicity that these loci are equivalent in their effect, that they act additively without dominance or epistasis, and that at each locus $P_1$ has two + alleles, each of which on average adds an amount $a$ to flower length, while $P_2$ has two − alleles, which have no effect. Thus $P_1$ has $2n$ + alleles and $P_2$ has $2n$ − alleles, so that $2na$ can be equated to the difference between the two parental means:

$$2na = \bar{P}_2 - \bar{P}_1 = 52.7.$$

Every $F_1$ plant has $n$ + alleles and $n$ − alleles, so that the mean value in $F_1$ should be the average of the two parental means, which is approximately true. In $F_2$ the number of + alleles in different plants will follow a binomial distribution with probability $\frac{1}{2}$ and index $2n$, provided that the loci are unlinked and so segregate independently of each other. Thus the mean value in $F_2$ should be the same as in $F_1$ (nearly but not quite true), whereas the variance should increase by an amount $\frac{1}{2}na^2$ because of the binomial variability in the number of + alleles between plants. Estimating the environmental variance in $F_2$ as 5.8 (calculated as a weighted average of the variances in $P_1$, $P_2$, and $F_1$ with $F_1$ receiving double weight), we obtain the equation

$$\tfrac{1}{2}na^2 = 41.6 - 5.8 = 35.8.$$

Solving these two equations we obtain the estimates

$$n = 9.7$$

$$a = 2.7.$$

East also gives the distribution of flower length in nine $F_3$ families derived from nine $F_2$ plants with different flower lengths. The means and variances of these distributions are shown in Table 1.4. There is a close relationship between the mean of an $F_3$ family and the flower length in its $F_2$ parent, which confirms that much of the variability in the $F_2$ plants is of genetic origin. (By contrast all $F_2$ families should have the same distribution regardless of the flower length of their $F_1$ parent because all the variability in $F_1$ is of environmental origin.) The variance within $F_3$ families is considerably smaller than the $F_2$ variance, but larger than the $F_1$ or the parental variances, as expected as a result of the progressive loss of genetic variability under continued selfing.

Thus the main features of East's results can be explained by supposing that the difference in flower length between the two varieties is of genetic origin and is controlled by about ten loci, each of which has a rather small effect. It can easily be seen how an effectively continuous, unimodal distribution is produced in $F_2$ under this model when a small amount of environmental variability is superimposed, in contrast to the non-overlapping classes typical of single gene inheritance. In consequence the statistical analysis of such polygenic characters

TABLE 1.4

*Mean and variance of flower length in nine $F_3$ families*

| Flower length in $F_2$ parent | Mean of $F_3$ family | Variance of $F_3$ family |
|---|---|---|
| 46 | 53.5 | 13.9 |
| 50 | 50.2 | 10.0 |
| 50 | 53.0 | 9.2 |
| 60 | 56.3 | 16.5 |
| 72 | 73.1 | 14.5 |
| 77 | 73.0 | 24.9 |
| 80 | 74.0 | 23.4 |
| 81 | 76.3 | 25.5 |
| 82 | 80.2 | 22.6 |

relies heavily on the calculation of means, variances and similar quantities used to characterize continuous frequency distributions. The above model is of course based on several simplifying assumptions, some of which will be relaxed in the more detailed discussion in Chapter 5.

## Outbred populations

Experiments like the above presuppose the existence of homozygous true-breeding varieties (pure lines) between which crosses can be made. The occurrence of such pure lines is confined to naturally self-fertilizing plants and to artificial populations on which selfing, brother–sister mating or some other system of close inbreeding has been imposed by the breeder. In natural populations of many plants and most animals outbreeding is ensured either by the existence of separate sexes or by mechanical, physiological or genetic devices which prevent self-fertilization. An outbred population forms a single reproductive unit in which the genes are reshuffled each generation by recombination and segregation. Its structure is in marked contrast to an inbred population broken up into a large number of reproductively isolated pure lines.

The analysis of genetic variability of quantitative characters in outbred populations is based on observing correlations between relatives. The statistical theory of correlation and regression arose from Francis Galton's work on the inheritance of human stature shown in Fig. 1.4. Galton collected data on

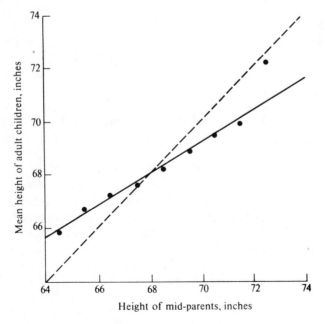

Fig. 1.4. Regression of height of adult children on their mid-parents (Galton 1887).

the heights of adult children and their parents in 205 families in 1884; after multiplying the female heights by 1.08 to make them comparable with male heights, he considered the relationship between the mean height of the children and the average height of their parents (their mid-parent). He found that there was a strong association between the heights of parents and children, but that the average deviation from the mean among the children was less than the corresponding deviation in their mid-parents; this is reflected by the differences between the observed points and the dotted line with unit slope in Fig. 1.4. Galton expressed this by saying that the children showed a 'regression towards mediocrity', and he estimated this filial regression (the slope of the solid line through the points) as 2/3–'that is to say the proportion in which the son is, on average, less exceptional than his mid-parent'. Galton was of course unaware of Mendel's laws. By a chain of rather dubious arguments he inferred from the fact of regression towards the mean his law of ancestral inheritance which he stated as follows: 'The two parents contribute between them on average one-half of the total heritage of the offspring, the four grandparents one-quarter, the eight grandparents one-eighth, and so on.'

Yule (1902) pointed out the ambiguity of this law and suggested a simpler explanation of regression towards the mean. Suppose that an individual's actual height is determined partly by his genotype and partly by the environment; we write $Y = G + E$, where $Y$ is the observed phenotypic value, $G$ is the genotypic value (the mean value among all individuals with the same genotype), and $E$ is a deviation due to environmental factors. Now an individual may have a high phenotypic value through having either a high genotypic value ($G$) or a high environmental deviation ($E$). Thus among parents with high phenotypic values, the genotypic value will on average be increased by a rather smaller amount than the phenotypic value, and there will be regression towards the mean among their offspring. We shall see later that if genes act additively the regression of child on mid-parent is a measure of the proportion of the phenotypic variance which is of genetic origin. Non-additive gene action (due to dominance or epistasis) causes a further reduction in the slope of the regression because only part of the genotypic value of the mid-parent is transmitted to the offspring.

Galton's important contribution was the invention of the concepts of regression and correlation, whose mathematical theory was developed by the early biometricians, in particular Karl Pearson. His law of ancestral inheritance was a stumbling block to progress since it led to the futile controversy between the biometricians and the Mendelians. One of the main causes of this controversy was that the early biometricians tended to regard Galton's work as an alternative theoretical framework for quantitative genetics rather than as an empirical description of the facts which must be interpreted in the light of Mendelism. The latter viewpoint only came to prevail gradually.

Yule suggested as early as 1902 that a Mendelian explanation of Galton's work was possible. In 1904 Karl Pearson worked out a detailed theory for a quantitative character determined by an arbitrary number of loci each with two equally frequent alleles of which one was completely dominant to the other. He first showed that on the assumption of random mating the frequencies of the three genotypes $AA$, $Aa$, and $aa$ at each locus would retain the constant values of $\frac{1}{4}$, $\frac{1}{2}$, and $\frac{1}{4}$ from one generation to the next. This is a special case of the law which bears the names of Hardy and Weinberg, who stated it independently in 1908, but which had already been foreshadowed by Yule (1902) and Castle (1903). Karl Pearson then derived the predicted regressions between relatives under the above model. He showed that the theoretical regression of offspring on one parent was a straight line with a slope of 1/3, and that the regression of offspring on mid-parent was a hyperbola, which tended to a straight line with a slope of 2/3 when the number of loci was large. He pointed out that these theoretical values were close to the actual values obtained by Galton, but that subsequent work had shown that the observed correlations were higher than this, and also varied from character to character. He concluded that the theory 'was not sufficiently elastic to cover the observed facts', but he did not attempt to generalize it since he considered complete dominance to be an essential feature of Mendelism. Yule (1906) pointed out that sufficient flexibility could be introduced into the theory by allowing dominance to be incomplete or absent and by introducing a term for the effect of the environment on the phenotype, but the precise formulation of the general theory did not come until 1918, when Fisher published his paper on 'The correlation between relatives on the supposition of Mendelian inheritance'. We shall return to this subject in Chapters 6 and 8. In the next chapter we shall consider in more detail the roles of genotype and environment in determining the observed phenotypic value.

**Problems**

1. When red-flowered four-o'clock plants are crossed to white-flowered plants, all the progeny have pink flowers; when pink-flowered plants are selfed, red-, pink-, and white-flowered plants appear among their progeny in the ratio 1:2:1. What will happen if pink-flowered plants are crossed to (a) red-flowered, (b) white-flowered plants?

2. Presence or absence of horns in cattle is a simple Mendelian trait, the hornless condition being dominant over horned. A hornless bull is mated to a horned cow. What can you infer about the genotype of the bull if the calf is (a) horned, (b) hornless?

3. In the snapdragon, a plant with white flowers of normal shape was crossed to a plant with red flowers of abnormal shape. The progeny all had pink flowers of normal shape, and when they were allowed to self-fertilize the following results were obtained in the next generation:

|  | Red | Pink | White |
| --- | --- | --- | --- |
| Normal shape | 39 | 94 | 45 |
| Abnormal shape | 15 | 28 | 13 |

## 16 The biological background

(Baur 1907; quoted by Strickberger 1976). Show that colour and shape are inherited independently of each other, normal being dominant to abnormal shape and pink-flowered plants being heterozygous for colour.

4. Derive the predicted $27:9:28$ ratio on page 4. What will happen if the purple-flowered $F_1$ plants are crossed back to the white-flowered parental varieties?

5. When a purple-flowered variety of sweet pea with long pollen was crossed to a red-flowered variety with round pollen, all the $F_1$ progeny had purple flowers and long pollen. When they were allowed to self-fertilize the following results were obtained in $F_2$:

|              | Purple | Red  |
| ------------ | ------ | ---- |
| Long pollen  | 4831   | 393  |
| Round pollen | 390    | 1338 |

(Punnett 1917; quoted by Whitehouse 1973). Show that the marginal distributions in $F_2$ agree with the predicted $3:1$ ratios for dominant genes affecting flower colour and pollen shape, but that these genes are not segregating independently of each other. Suppose that they are on the same chromosome with recombination fraction $r$. Evaluate the Expected proportions of the four phenotypes in $F_2$, and show that there is good agreement between observed and predicted numbers if $r$ is taken as 0.12. (See Bailey 1961 for statistical methods of estimating $r$.)

6. White eye colour in *Drosophila* is due to a sex-linked gene which is recessive to the normal red eye colour in females. If a white-eyed male is crossed to a homozygous red-eyed female, what will be the eye colour of their offspring? If brothers and sisters from this cross are mated together what will be the frequency of red eyes (a) among their daughters, (b) among their sons?

7. If a black tomcat mates a tortoiseshell female (black and ginger mosaic), what coat colours would you expect in the kittens?

8. Suppose that $F_1$ plants in Table 1.3 are crossed back to the parental line $P_2$. What will be the distribution of the number of + alleles in the backcross under the hypothesis suggested? Hence find the predicted mean and variance in the backcross.

9. Plot the $F_3$ family mean against the parental value in Table 1.4 and calculate the slope of the regresion. Explain why you would expect to find regression towards the mean in this situation.

# 2 Genotype and environment

*Most quantitative characters are affected by both genetic and environmental factors. In this chapter we shall discuss the roles of genotype and environment in determining the observed phenotypic value of a character. We begin by considering in more detail the distinction between genotypic and phenotypic values, and we shall then discuss genotype–environment interaction*

## Phenotypic and genotypic values

We shall first consider the classical experiments of the Danish biologist Johannsen at the beginning of this century which clarified the important distinction between the genetic and the environmental contributions to the observed value of a character. Johannsen (1903, 1909) studied the weight of seeds in a commercial variety of the haricot bean; in all his experiments he allowed the plants to self-fertilize naturally. He first showed that some of the variability in seed weight is of genetic origin since there is a positive relationship between the seed weights of parent and offspring (Table 2.1). He then isolated 19 pure lines, each derived by continued selfing from a single plant, and showed that there was no longer a correlation between parents and offspring belonging to the same line; thus the within-line variability must be entirely of environmental origin. He confirmed this point by showing that selection for seed weight within a pure line was ineffective. Table 2.2 shows that there is a consistent difference between two pure lines and some variability in seed weight from year to year, but that there is no difference between two sublines within a

TABLE 2.1

*Parent–offspring correlation in a mixed population and in a single pure line of haricot beans in 1902 (Johannsen 1903)*

| Seed weight of parent (mg) | Average seed weight of offspring (mg) | |
|---|---|---|
| | Total population | Single pure line |
| 150– | 440 | |
| 250– | 443 | 475 |
| 350– | 461 | 450 |
| 450– | 490 | 451 |
| 550– | 519 | 458 |
| 650– | 560 | |
| Parent–offspring correlation | 0.34 | −0.02 |

TABLE 2.2

*The effect of selection within two pure lines of beans (Johannsen 1909)*

| Year | Mean seed weight of offspring (mg) | | | |
|---|---|---|---|---|
| | Line 1 | | Line 19 | |
| | Low subline | High subline | Low subline | High subline |
| 1902 | 632 | 648 | 358 | 348 |
| 1903 | 752 | 709 | 402 | 410 |
| 1904 | 546 | 567 | 314 | 326 |
| 1905 | 636 | 636 | 383 | 392 |
| 1906 | 744 | 730 | 379 | 399 |
| 1907 | 691 | 677 | 374 | 370 |

line selected for low and high seed weight respectively. (The low subline was maintained by sowing the smallest seeds of that subline to produce next year's plants, and the high subline by sowing the largest seeds.)

These results demonstrate the fact that in a natural population of a self-fertilizing species all individuals will be homozygous at all loci, though different individuals may be homozygous for different alleles. The population is thus subdivided into a number of reproductively isolated pure lines, the members of each line being genetically identical but different from the members of any other line. Thus variability within a line is due to non-genetic, environmental causes; it is not inherited by the offspring and does not respond to selection. In contrast, variability between the mean values of different lines is entirely of genetic origin and does respond to selection.

Following Johannsen's terminology we distinguish between the *phenotypic value*, the value of a character (such as seed weight or flower length) actually observed, and the *genotypic value*, the mean value in a large group of individuals with the same genotype raised under similar conditions (for example, in the same place in the same year). We denote the phenotypic value by $Y$, the genotypic value by $G$, and the difference between them, usually called the environmental deviation, by $E$. We may therefore write

$$Y = G + E. \tag{2.1}$$

It is only possible to observe a large number of individuals with the same genotype if we have a pure line, or an $F_1$ cross between two pure lines, or if vegetative propagation is possible. In other cases the large group of genetically identical individuals is hypothetical; in man, for example, the nearest we can get is a pair of identical twins.

The environmental deviation, $E$, represents differences between individuals of the same genotype raised under similar conditions (for example, in the same year and location). It will be assumed that any environmental variability not controlled by the experimenter has been randomized by allocating genotypes at random to plots or other experimental units; in this way one can ensure that there are no systematic differences between the environmental factors affecting different genotypes. The environmental deviation reflects two sources of variability (in addition to errors of measurement which are usually negligible): (i) differences in the external environment, for example, fertility differences between plots; (ii) variability in the internal environment of the organism, that is to say random variability in development. Developmental variability may be of considerable importance in some characters. For example, Mather and Jinks (1977) counted the sternopleural bristles on both sides of a large number of flies from an inbred line of *Drosophila*. The variance of the total bristle number (the sum of the two sides) was 2.20, while the variance of the difference between the two sides was 2.00. They concluded that 91 per cent of the variance of total bristle number (2.0/2.2) was due to factors which affected the two sides of the same fly independently. Such factors are likely to be local developmental accidents rather than variations in the external environment which would affect both sides the same. (Let bristle number on left side $= X_L + Z_L$, number on right side $= X_R + Z_R$, where $X_L$ and $X_R$ are independent random variables whereas $Z_L = Z_R$; thus the $X$s represent factors which affect the two sides independently, while the $Z$s represent factors which affect the two sides in the same way. Then

$$\text{Var(Sum)} = 2 \, \text{Var}(X) + 4 \, \text{Var} \, (Z)$$

$$\text{Var(difference)} = 2 \, \text{Var} \, (X).)$$

Whatever its origin, the environmental deviation can be regarded as a random error with zero mean. In the simplest case the distribution of $E$, and in particular its variance, will be the same for all genotypes, but if some genotypes are more sensitive than others to environmental change then the variance of $E$ will vary with the genotype. In particular Lerner (1954, 1958) has argued that inbred lines derived from normally outbreeding species have an increased environmental variance because homozygotes are less well buffered than heterozygotes against environmental fluctuations; he gives evidence that inbred lines have a higher variance in some cases than the $F_1$ cross between them. For further discussion see Falconer (1960, pp. 270–2) and Eanes (1978).

Heterogeneity of the error variance can sometimes by removed by an appropriate transformation. For example, the two lines in Table 2.2 had standard deviations of 110 and 65 mg respectively in 1902, the means being 642 and 351 mg; the corresponding coefficients of variation are 17.1 and 18.5 per cent. The constancy of the coefficient of variation suggests that a logarithmic transformation will stabilize the variance.† When the error variance

---

† A variance stabilizing transformation is a transformation which makes the variance constant and independent of the mean. There is a statistical rule that if the standard deviation is proportional to the mean (constant coefficient of variation), then a logarithmic transformation will make the variance constant; if the variance is proportional to the mean (constant variance/mean ratio), a square-root transformation will achieve this end.

can be stabilized by a simple scale transformation it is desirable to do so, though it is more important to find a transformation to remove genotype–environment interaction and epistasis when this is possible; if the experimenter is lucky the same transformation will achieve all three objectives. If the error variance is a function of genetic heterozygosity it will not usually be possible to find a variance-stabilizing transformation.

If different genotypes occur with different frequencies in the population (which may be either a mixture of pure lines or an outbreeding population), the genotypic value $G$ is a random variable with a distribution determined by the frequencies of the genotypes. In the simple case in which the distribution of $E$ is the same for all genotypes, $G$ and $E$ have independent distributions. If the distribution of $E$ depends on the genotype, then $G$ and $E$ will not be independent but they will be uncorrelated since the Expected value of $E$ is by definition zero for all genotypes. Thus

$$\mathrm{Var}(Y) = \mathrm{Var}(G) + \mathrm{Var}(E)$$

or                                                                                                       (2.2)

$$V_Y = V_G + V_E.$$

This decomposition of the phenotypic variance into genetic and environmental components is of great importance. The genetic variance, $V_G$, is the variance of the genotypic values, $G$, in the population. The environmental variance, $V_E$, is the average variance of $E$ over all genotypes. Thus $V_E$ may change if the genotypic constitution of the population changes. However, it will often be assumed for simplicity that $V_E$ is constant.

### Genotype–environment interaction

The genotypic value was defined in the last section as the mean value in a group of individuals with the same genotype raised under similar conditions. It was assumed that any environmental variation not controlled by the experimenter was randomized; this *micro-environmental* variation arises from small-scale fluctuations in environmental conditions. The experimenter may also wish to study the effect of large changes in environmental conditions (*macro-environmental* variation). The different macro-environmental conditions may be experimental treatments controlled by the experimenter, such as differences in temperature or food supply, or they may be different years or places in experiments repeated in time or space.

The model (2.1) can be extended to situations in which the genotypes have been tested under several different macro-environmental conditions. Write $Y_{ijk}$ for the value of the $k$th individual of the $i$th genotype under the $j$th condition. The statistical model for this crossed classification is

$$Y_{ijk} = m + G_i + C_j + I_{ij} + E_{ijk}.$$                                          (2.3)

In this representation $m$ is the mean value over all genotypes and conditions, $G_i$ is the effect of the $i$th genotype (averaged over conditions and represented as a deviation from the mean), $C_j$ is the effect of the $j$th condition (averaged over genotypes and represented as a deviation from the mean), $I_{ij}$ is the genotype–condition interaction (representing any failure of these two factors to act additively, and usually called the genotype–environment interaction), and $E_{ijk}$ is the environmental deviation of the $k$th individual within the $ij$th genotype–condition. The micro-environmental deviations $E_{ijk}$ are by definition independent random variables with zero mean, but their distribution may depend on $i$ and $j$, that is to say upon the genotype and the environmental condition. An analysis of variance will decompose the variance into terms for genotypes, environmental conditions, genotype–environment interaction, and micro-environmental variation. The term for genotype–environment interaction is of particular interest because it reveals whether there is a differential response of different genotypes to environmental changes.

As an example, Table 2.3 shows the analysis of variance on the growth rate of 38 races of *Arabidopsis* (a self-fertilizing plant) grown at six temperatures; ten plants of each race were grown at each temperature (Griffing and Langridge 1963). The Expected mean squares are given on the assumption that the races represent a random sample of homozygous forms of the species (random effects model), whereas temperatures represent a fixed set of environmental conditions (fixed effects model). All the effects are overwhelmingly significant. The components of variance were estimated by equating observed and Expected mean squares. The interaction variance, $V_I$, is about equal to the genotypic variance, $V_G$, so that there is considerable variation in the way in which different races react to temperature.

A useful measure of the importance of genotype–environment interaction is provided by the ratio $V_G/(V_G + V_I)$, which can be interpreted as the corre-

TABLE 2.3

*Analysis of variance of growth rate of 38 races of* Arabidopsis *grown at six temperatures (modified from Griffing and Langridge 1963)*

| Source of variation | Degrees of freedom | Mean square | Expected mean square |
|---|---|---|---|
| Races (genotypes) | 37 | 4.915 | $60V_G + V_E$ |
| Temperatures (conditions) | 5 | 366.784 | $380V_C + 10V_I + V_E$ |
| Races × temperatures | 185 | 0.956 | $10V_I + V_E$ |
| Micro-environment | 2052 | 0.110 | $V_E$ |

$\hat{V}_G = 0.080$, $\hat{V}_C = 0.963$, $\hat{V}_I = 0.085$, $\hat{V}_E = 0.110$.
Note: $G$, $I$, and $E$ are random effects with corresponding variances, $C$ is a fixed effect with $V_C = \sum C_j^2/5$. $V_I$ occurs in the Expected mean square for temperatures but not races because the latter are random while the former are fixed effects. There is some controversy about the analysis of the mixed model, with some factors fixed and others random. I follow here the treatment of Scheffé (1959, Chapter 8) and Snedecor and Cochran (1969, Chapter 12).

lation between the values of individuals with the same genotype raised in different macro-environments. From the data of Table 2.3 this correlation can be estimated as about 0.48.

A similar experiment can be done to investigate the interaction between breeds or races of an outbred population with their environment. In this case it should be remembered that the breeds represent different mixtures of genotypes, not single genotypes; the residual error variance, $V_E$, contains genetic variation within races as well as micro-environmental variation, and the variance between races, $V_G$, represents genetic variation between races.

A common experiment compares the behaviour of different varieties in a number of different places over several years, allowing interactions in both space and time to be measured. The model underlying the analysis may be written

$$Y_{ijkl} = m + G_i + P_j + T_k + (GP)_{ij} + (GT)_{ik} + (PT)_{jk} + (GPT)_{ijk} + E_{ijkl} \quad (2.4)$$

In this model $Y_{ijkl}$ is the $l$th observation on the $i$th variety in the $j$th place in the $k$th year. The first four terms on the right hand side are the mean and the main effects of genotypes (varieties), places, and times (years); the next three terms are the first order interactions, then the second order interaction, and finally the micro-environmental deviation within places and years. It is usually assumed, perhaps rather artificially, that varieties, places and years can all be regarded as random samples chosen from large populations, so that the model is a random effects model. The analysis of variance is shown algebraically in Table 2.4, on the assumption that $v$ varieties have been tested in $p$ places for $t$ years, with $r$ replications. The variance components can be estimated by equating observed and Expected values. Standard errors can be placed on the estimated variance components by using the fact that under normality the mean squares are independent chi-square variates, apart from a scale factor. If $M$ is a mean square with $f$ degrees of freedom, then $\mathrm{Var}(M) = 2E^2(M)/f$; an unbiased estimate of $\mathrm{Var}(M)$ is given by $2M^2/(f + 2)$. (Note: this is only valid for the random effects model.)

TABLE 2.4

*General form of analysis of variance for year and place interactions*

| Source of variation | Degrees of freedom | Expected mean square |
|---|---|---|
| Varieties | $(v - 1)$ | $rptV_G + rtV_{GP} + rpV_{GT} + rV_{GPT} + V_E$ |
| Places | $(p - 1)$ | $rvtV_P + rtV_{GP} + rvV_{PT} + rV_{GPT} + V_E$ |
| Years | $(t - 1)$ | $rvpV_T + rpV_{GT} + rvV_{PT} + rV_{GPT} + V_E$ |
| Varieties × places | $(v - 1)(p - 1)$ | $rtV_{GP} + rV_{GPT} + V_E$ |
| Varieties × years | $(v - 1)(t - 1)$ | $rpV_{GT} + rV_{GPT} + V_E$ |
| Places × years | $(p - 1)(t - 1)$ | $rvV_{PT} + rV_{GPT} + V_E$ |
| Varieties × places × years | $(v - 1)(p - 1)(t - 1)$ | $rV_{GPT} + V_E$ |
| Residual | $vpt(r - 1)$ | $V_E$ |

Estimates of genotype and genotype–environment interaction variances from four experiments are shown in Table 2.5. A general feature of such analyses in agricultural crops is the absence or weakness of the variety × place interaction, and the common (though not universal) presence of the second order variety × place × year interaction. The presence of a variety × place interaction would suggest to the breeder the existence of varieties specially adapted to different localities; the absence of this type of interaction is probably due to the fact that previous selection of varieties to suit localities has removed the interaction. The presence of interaction with years cannot be used to select for varieties adapted to particular years since the environment of future years cannot be predicted in advance.

TABLE 2.5

*Estimates of genotype-interaction variances*

| Component | Cotton[1] Lint yield | Tobacco[2] Leaf yield | Sheep[3] Wool weight | Barley[4] yield |
|---|---|---|---|---|
| $V_G$ | 0.028 | 40 719 | 1.701 | 15.02 |
| $V_{GP}$ | 0.002 | 100 | 0.045 | 0.02 |
| $V_{GT}$ | 0.001 | 1990 | 0 | 3.99* |
| $V_{GPT}$ | 0.016* | 7002* | 0.074* | 15.97* |
| $V_E$ | 0.063 | 20 913 | 0.829 | 42.78 |

* Significant at 5 per cent level or better; unstarred interactions are not significant. Significance tests were done on the original analysis of variance.
[1] 15 varieties at nine places in North Carolina over three years (Miller, Williams, and Robinson 1959).
[2] Seven varieties at five places over three years (Jones, Matzinger, and Collins 1960).
[3] Five strains of Merino at three places in Australia over four years (Dunlop 1962).
[4] Six varieties at nine places in Minnesota over four years (Rasmusson and Lambert 1961).

It is important to remember that the genotypic value is by definition an average value over a range of conditions and therefore depends on the conditions over which the genotypes have been tested. In particular, if measurements are confined to a particular year and place, the genotypic value in that year–place will include all the terms in eqn (2.4) except the micro-environmental deviation, and the genotypic variance will include contributions which would be counted as genotype-interaction variances if observations over several years and places were made. Thus observations in a single year and place will overestimate the 'true' genotypic variance which would be estimated from observations over several years and places.

In the above account we have ignored a number of complications in the experimental design which often arise in practice and which must be taken into account in the analysis. In each place the experiment may be laid out in a randomized blocks design, allowing a variety × blocks interaction to be measured. If the experiment is repeated over several years with the same blocking, the analysis must take into account the fact that blocks form a nested classification within places, whereas places, varieties and years are all crossed factors. Another complication is that several plants of the same variety may be grown in each plot, in addition to replication of varieties on different plots in each block. In this situation we must take into account the existence of two sources of micro-environmental error, within plots and between plots; see Table 5.4 (p. 66) for an example. In situations when the variability between plants in the same plot is not of interest, this complication may be avoided by taking the plot mean as the unit of observation.

So far we have considered the interaction which can be measured when genetically different groups of plants or animals are raised under different environmental conditions. The different groups may be homozygous pure lines, or they may be different breeds or races of an outbreeding species. In the latter case the breeds are themselves genetically heterogeneous, but they differ in their mean genotypic values and we are considering how these mean values behave in different environments. We are thus concerned with the variability between breeds and between environments and the interaction between the two; the genetic variability within breeds is treated as part of the residual error.

It may also be important to consider the interaction between different genotypes within a breed and the environment. As an example, progeny tests of four bulls measured in an experimental station and on farms are shown below (Pirchner 1969); each progeny test is the average milk yield (in kg) of a large number of the bull's daughters:

| Bull | Experimental station | Farms |
|------|---------------------|-------|
| A    | 4380                | 3760  |
| B    | 4350                | 3560  |
| C    | 4130                | 3880  |
| D    | 4060                | 3640  |

There is an environmental difference; milk yield is higher at the experimental station than on the farms, presumably because of better management. There is also genotype–environment interaction: bull A was the best bull as judged at the experimental station, but bull C was the best as judged on the farm. This is important to animal breeders since it implies that selection of bulls for breeding based on the performance of their daughters at experimental stations may not select the best bulls for use on farms. We shall consider this type of genotype–environment interaction more fully in Chapter 6.

The estimation of variance components provides an overall measure of the importance of genotype–environment interaction, but gives no indication of its nature. The first step in a more detailed examination is to consider whether the interaction is an artefact of the scale on which the observations are measured, which can be removed by a simple transformation such as taking logarithms or square roots. Interactions which can be removed by a scale transformation are of little biological importance and should be removed since this simplifies both the analysis and the interpretation of the data. For example, after averaging over sublines in Table 2.2 it will be found that there are large differences in seed weight between the two lines of beans and between the six years, but that there is also some genotype–year interaction; the difference between the lines is larger in good years than in bad years. This interaction is effectively removed by a logarithmic transformation, which will also stabilize the variance as we found earlier.

Genotype–environment interaction can sometimes be removed by a scale transformation, but it often represents a real biological fact which must be taken into account. For example, Falconer (1960) gives the following data on the growth of two inbred strains of mice reared on two levels of nutrition:

|  | Good nutrition | Bad nutrition |
|---|---|---|
| Strain A | 17.2 | 12.6 |
| Strain B | 16.6 | 13.3 |

As expected, the level of nutrition has a big effect on the growth rate (measured as the change in weight in grams between three and six weeks of age) but there is also an important genotype–environment interaction: strain A grows better than strain B under good conditions but worse under bad conditions. This type of interaction is clearly of biological importance. We may describe it by saying that strain B is more stable to nutritional change than strain A.

### *Stability of genotypes

We shall now consider attempts to describe interactions in terms of genotypes of different 'stabilities' to environmental change. Suppose that $v$ varieties have been tested in $c$ macro-environmental conditions with $r$ replications in each, and write the model:

$$Y_{ijk} = m + G_i + C_j + I_{ij} + E_{ijk}. \qquad (2.3 \; bis)$$

(If the environments are years and places, we suppose that each year–place is treated as a separate environment, the factorial structure being ignored.) Let us now concentrate on a particular variety, say the $i$th variety. For fixed $i$, we write the model

$$Y_{ijk} = m + G_i + C_j(i) + E_{ijk} \qquad (2.5)$$

where $C_j(i) = (C_j + I_{ij})$ is the macro-environmental effect of the $j$th condition on the $i$th variety. The analysis of variance for this variety may be written:

| Source of variation | Mean square | Expected mean square |
|---|---|---|
| Between environments | $M_B = r \sum_j (Y_{ij.} - Y_{...})^2/(c-1)$ | $rV_C(i) + V_E(i)$ |
| Within environments | $M_W = \sum_{j,k} (Y_{ijk} - Y_{ij.})^2/c(r-1)$ | $V_E(i)$ |

In this analysis $V_C(i) = \text{Var}(C_j(i))$ and $V_E(i)$ is the variance of $E_{ijk}$ with $i$ fixed. We can estimate $\hat{V}_C(i) = (M_B - M_W)/r$ and $\hat{V}_E(i) = M_W$. $V_C(i)$ is a measure of the stability of the $i$th genotype to macro-environmental change, and $V_E(i)$ is a measure of its stability under micro-environmental change, a small variance meaning stability and a large variance instability. Griffing and Langridge (1963) used these measures to compare the stability of homozygous inbred lines and of heterozygous $F_1$ and $F_2$ crosses derived from them in the self-fertilizing plant *Arabidopsis* grown at different temperatures. They found a reduction in $V_C(i)$ in heterozygotes compared with homozygotes, but no change in $V_E(i)$. (Compare Table 2.3. $V_C(i)$ but not $V_E(i)$ can be estimated for $F_2$ which is a mixture of genotypes.)

This line of thinking can be taken a stage further. If interaction is due to differences in stability between genotypes, we might expect that all genotypes would react in the same direction but by different amounts in response to environmental changes. In other words we would expect $C_j(i)$ to be proportional to $C_j$, say $C_j(i) = \beta_i C_j$, where $\beta_i$ is a measure of the stability of the genotype; $\beta_i$ between 0 and 1 would indicate a stable genotype while $\beta_i$ above 1 would indicate an unstable genotype. We therefore suppose that

$$C_j(i) = \beta_i C_j + I_{ij}^* \tag{2.6}$$

where $I_{ij}^*$ is the residual interaction after allowing for differences in stability between genotypes. The full model may thus be written

$$Y_{ijk} = m + G_i + \beta_i C_j + I_{ij}^* + E_{ijk}. \tag{2.7}$$

If the model is successful then most of the interaction variance will be 'explained' by fitting the stability parameters $\beta_i$.

The model (2.7) is non-linear, and some complications arise in its statistical analysis. The following approximate procedure is usually adopted. First analyse the model (2.3), obtaining among other things the estimates

$$\hat{C}_j = Y_{.j.} - Y_{...}$$
$$\hat{I}_{ij} = Y_{ij.} - Y_{i..} - Y_{.j.} + Y_{....}. \tag{2.8}$$

Substituting $\hat{C}_j$ for $C_j$ in (2.7) we obtain the estimates

$$\hat{\beta}_i = \sum_j Y_{ij.} \hat{C}_j / \sum_j \hat{C}_j^2$$
$$\hat{I}_{ij}^* = Y_{ij.} - Y_{i..} - \hat{\beta}_i \hat{C}_j = \hat{I}_{ij} - (\hat{\beta}_i - 1)\hat{C}_j. \tag{2.9}$$

The component of the interaction sum of squares accounted for by this procedure (the difference between $r\Sigma\hat{I}_{ij}^2$ and $r\Sigma\hat{I}_{ij}^{*2}$) may be taken as having $(v-1)$ degrees of freedom. (The heuristic argument is that $v$ stabilities have been estimated but that one degree of freedom is lost since their average value is constrained to be 1. The procedure can be regarded as the first round of an iterative estimation by non-linear least squares (Draper and Smith 1966, Chapter 10).)

TABLE 2.6

*Analysis of variance of stability (Finlay and Wilkinson 1963)*

| Source | Degrees of freedom | Mean square |
|---|---|---|
| Genotypes | 276 | 0.5618 |
| Environments | 6 | 125.5803 |
| Genotypes × environments | 1656 | 0.0616 |
| Regressions | 276 | 0.2227 |
| Deviations from regressions | 1380 | 0.0294 |
| Residual | 3864 | 0.0186 |

An analysis of this sort was done by Finlay and Wilkinson (1963) on the log yield of 277 varieties of barley in seven environments in Australia (two places over three years and a third place for one year only). The analysis of variance is shown in Table 2.6. All mean squares are significant, but it will be seen that regression of genotype on environment accounts for a large proportion of the interaction variance. Thus a lot of the interaction is due to differences in stability between genotypes, which can usefully be described by the regression coefficients. For example, the variety Bankuti Korai is very stable, with $\hat{\beta} = 0.14$; it therefore produces higher than average yields in poor environments and lower than average in good environments. By contrast the variety Provost is very sensitive to environmental change, with $\hat{\beta} = 2.13$, and is thus very productive under favourable conditions, but quickly becomes unproductive as the environment deteriorates.

This simple method of analysis circumvents the non-linear nature of the model (2.7), but it seems to work well in practice and has been found valuable by plant breeders. The variability of the residual interactions, $I_{ij}^*$, for fixed $i$ can be used as another measure of the stability of the $i$th variety, measuring a different aspect of stability (Eberhart and Russell 1966). An estimate of this variance is given by

$$\sum_j \hat{I}_{ij}^{*2}/(c-1) - \hat{V}_E/r.$$

## Problems

1. The mean seed-weights of Johannsen's 19 pure lines in 1901 and 1902 and the standard deviations in 1902 are shown below (data in mg):

```
1901 (mean) 600 520 570 600 512 395 440 405 395 400
1902 (mean) 642 558 554 547 512 ·506 492 488 482 465
1902 (SD)    109  93  76  84  76   64  69  72  76  79

1901 (mean) 380 410 400 390 510 360 340 312 310
1902 (mean) 455 455 454 453 450 446 428 407 351
1902 (SD)    70  66  74  75  66  69  72  78  65
```

Plot the means and find the regression of 1902 weights on 1901 weights. What explanations are possible for the regression towards the mean?

2. If the 19 lines were mixed together in equal numbers in 1902, find $V_Y$, $V_G$, and $V_E$.

3. Mather and Jinks (1977) present the following data on the mean numbers of sterno-pleural bristles in two inbred lines of *Drosophila* ($P_1$ and $P_2$), the cross between them ($F_1$) and the offspring of this cross ($F_2$) when raised in six possible environments; these environments comprised all the possible combinations of two temperatures (18 and 25 °C) and three types of culture vessel (called B, Y, and U).

|       | 18 °C |       |       | 25 °C |       |       |
|-------|-------|-------|-------|-------|-------|-------|
|       | B     | Y     | U     | B     | Y     | U     |
| $P_1$ | 20.58 | 20.51 | 20.26 | 20.44 | 20.93 | 20.66 |
| $P_2$ | 19.63 | 19.34 | 19.34 | 18.67 | 18.14 | 17.61 |
| $F_1$ | 19.98 | 20.01 | 20.16 | 19.22 | 18.93 | 18.48 |
| $F_2$ | 20.19 | 19.86 | 19.75 | 19.45 | 18.68 | 18.75 |

The means for $P_1$ and $P_2$ are each based on five replicate cultures, those for $F_1$ on eight replicates, and those for $F_2$ on two replicates; each replicate contained a large number of flies. The error mean square between replicates was 0.1007 with 96 degrees of freedom.

Analyse these data as a $4 \times 3 \times 2$ factorial design with unequal numbers of replications for different genotypes.

4. Find estimates of the stabilities of the genotypes in the above experiment, allowing for the unequal numbers of replications for different genotypes.

# 3 Genotype frequencies

*Genetic information about quantitative characters in an inbred population is obtained by crossing lines, in an outbred population by measuring correlations between relatives. To interpret the results of these experimental observations it is necessary in both cases to understand the structure of genotype frequencies in the population, which is determined by the breeding system. In this chapter we shall consider in turn the genotype frequencies in an inbred population following a cross between two lines and in an outbred population under random mating. We shall then describe how the concept of identity by descent can be used to define measures of the degree of inbreeding in a population and of the genetic relationship between relatives in an outbred population. For a fuller treatment of these topics the reader is referred to a book on population genetics, such as Crow and Kimura (1970), Jacquard (1974), or Li (1955).*

## Line crosses

A simple genetic experiment will begin by crossing two parental varieties, $P_1$ and $P_2$, to produce a hybrid generation, $F_1$. It will be assumed that the parental varieties are pure lines, either occurring naturally in a self-fertilizing species or developed artificially by close inbreeding in a naturally outbreeding species. In a self-fertilizing species we may obtain the successive generations $F_2$, $F_3$, $F_4$, and so on by continued selfing; thus $F_2$ is obtained by selfing $F_1$, $F_3$ by selfing $F_2$, and so on. Self-fertilization is impossible in species with separate sexes (most animals and some plants) and in this case we may use continued sib-mating to obtain the successive generations which will be denoted $S_2$, $S_3$, and so on. $S_2$ is obtained by mating $F_1$ individuals at random (they are all sibs). $S_3$ is obtained by mating $S_2$ individuals at random (which is genetically equivalent to sib-mating), $S_4$ is obtained by mating sibs from $S_3$ at random, and so on. Another useful mating system is backcrossing to the two parental varieties. The backcross generations $B_1$ and $B_2$ are obtained by crossing $F_1$ individuals to $P_1$ and $P_2$ respectively. These backcross generations may then be selfed or they may be crossed back again to the parental varieties ($B_1 \times P_1$ and so on). Backcrossing from the $F_2$ generation to obtain the crosses $F_2 \times P_1$, $F_2 \times P_2$ and $F_2 \times F_1$ has also been found useful.

The purpose of these experiments is to obtain information from the means and variances in the different generations about the genetic determination of the character under study. To develop the theory of how the means and variances ought to behave under specific genetic models it is necessary to know the genotype frequencies in the different generations. Suppose that at a particular locus $P_1$ and $P_2$ have genotypes $AA$ and $aa$ respectively, so that $F_1$ has the genotype $Aa$. In subsequent generations we write $p_0, p_1$, and $p_2$ for the frequencies of individuals with 0, 1, and 2 $A$ alleles, that is to say for the probabilities of the genotypes $aa$, $Aa$, and $AA$. Putting this another way, we may regard the number of $A$ alleles present in an individual as a random variable, denoted by $Z$, and we define

$$p_z = \text{Prob}[Z = z], \quad z = 0, 1, 2. \tag{3.1}$$

We shall now derive these probabilities under the different mating systems just described.

Homozygotes breed true on selfing while heterozygotes produce the genotypes $AA$, $Aa$, and $aa$ among their offspring with probabilities $\frac{1}{4}$, $\frac{1}{2}$, and $\frac{1}{4}$. Thus the frequency of heterozygotes is halved each generation, whereas the frequencies of the two types of homozygotes increase equally. Starting from the initial condition that $p_1 = 1$ in $F_1$, it is clear that in $f_k$

$$
\begin{aligned}
p_1 &= (\tfrac{1}{2})^{k-1} \\
p_0 &= p_2 = \tfrac{1}{2} - (\tfrac{1}{2})^k.
\end{aligned}
\tag{3.2}
$$

In the backcross and related generations the genotype frequencies can be calculated from the gametic outputs of the parents. For example, consider the backcross $B_1 = F_1 \times P_1$. Half the gametes from $F_1$ are $A$ and the other half are $a$, while the gametes from $P_1$ are all $A$; thus the genotype frequencies in $B_1$ are half $AA$ and half $Aa$. Some other results are shown in Table 3.1.

TABLE 3.1

*Genotype frequencies under different mating systems*

| Generation | Genotype | | |
|---|---|---|---|
| | $AA$ | $Aa$ | $aa$ |
| $P_1(P_2^*)$ | 1 | — | — |
| $F_1$ | — | 1 | — |
| $F_2 = S_2 = S_3$ | 0.2500 | 0.5000 | 0.2500 |
| $F_3$ | 0.3750 | 0.2500 | 0.3750 |
| $F_4$ | 0.4375 | 0.1250 | 0.4375 |
| $S_4$ | 0.3125 | 0.3750 | 0.3125 |
| $S_5$ | 0.3438 | 0.3125 | 0.3438 |
| $B_1(B_2^*)$ | 0.5000 | 0.5000 | — |
| $B_1 \times P_1(B_2 \times P_2^*)$ | 0.7500 | 0.2500 | — |
| $B_1 \times P_2(B_2 \times P_1^*)$ | — | 0.7500 | 0.2500 |
| $B_1$ selfed ($B_2$ selfed*) | 0.6250 | 0.2500 | 0.1250 |
| $F_2 \times P_1(F_2 \times P_2^*)$ | 0.5000 | 0.5000 | — |
| $F_2 \times F_1$ | 0.2500 | 0.5000 | 0.2500 |

* For the starred crosses reverse the frequencies of $AA$ and $aa$.

Under sib-mating it is necessary to keep track of the frequencies of the different mating types in each generation. There are six mating types, which can be grouped into four classes:

(1) $AA \times AA$ or $aa \times aa$    (2) $AA \times aa$

(3) $AA \times Aa$ or $Aa \times aa$    (4) $Aa \times Aa$

We define the transition probability $t_{ij}$ as the probability of going from class $j$ to class $i$ in one generation of sib-mating. The matrix of these transition probabilities, denoted by $\mathbf{T} = [t_{ij}]$, is

$$\mathbf{T} = \begin{bmatrix} 1 & 0 & \frac{1}{4} & \frac{1}{8} \\ 0 & 0 & 0 & \frac{1}{8} \\ 0 & 0 & \frac{1}{2} & \frac{1}{2} \\ 0 & 1 & \frac{1}{4} & \frac{1}{4} \end{bmatrix} \tag{3.3}$$

For example, the third column is obtained by observing that the mating $AA \times Aa$ gives rise to $AA$ and $Aa$ offspring in equal numbers; if these sibs mate at random, one-quarter of the matings will belong to class 1, one-half to class 3, and one-quarter to class 4. The mating $Aa \times aa$ gives the same result.

Denote the probability of a mating of class $i$ in $S_k$ by $q_i(k)$, and write $\mathbf{q}(k)$ for the vector of these probabilities. They satisfy the recurrence relationship

$$\mathbf{q}(k + 1) = \mathbf{T}\mathbf{q}(k), \tag{3.4}$$

whose solution is

$$\mathbf{q}(k) = \mathbf{T}^k \mathbf{q}(0). \tag{3.5}$$

Note that $\mathbf{q}(0) = (0\ 1\ 0\ 0)'$ since the initial cross was a mating of class 2. The powers of $\mathbf{T}$ can be found by direct calculation, but it is more useful to evaluate them by finding the spectral resolution. (See Bailey 1964, Chapter 5 for a good account of this method.) The latent roots of $\mathbf{T}$ are 1, 0.25, 0.809, and $-0.309$. Since they are all simple roots the mating class frequencies can be expressed in the form

$$q_i(k) = a_i + b_i(0.25)^k + c_i(0.809)^k + d_i(-0.309)^k. \tag{3.6}$$

Since mating class 1 is an absorbing state and the rest are transient states, it follows that $a_1 = 1, a_2 = a_3 = a_4 = 0$. The constants $b_i$, $c_i$, and $d_i$ can be found by evaluating $q_i(k)$ directly from eqns (3.4) or (3.5) for $k = 0, 1, 2$, thus obtaining three linear equations in three unknowns. From symmetry, the frequencies of the two mating types in class 1 are the same, and likewise for class 3; thus the individual mating type frequencies in classes 1 and 3 can be obtained by halving the class frequencies. The genotype frequencies in $S_k$ are found to be

$$\begin{aligned} p_0 = p_2 &= \tfrac{1}{2}q_1 + \tfrac{1}{2}q_2 + \tfrac{1}{4}q_3 = 0.5 - 0.447(0.809)^k + 0.447(-0.309)^k \\ p_1 &= \tfrac{1}{2}q_3 + q_4 = \qquad\qquad 0.894(0.809)^k - 0.894(-0.309)^k \end{aligned} \tag{3.7}$$

The above results show the probabilities with which an individual in a specified generation will have the genotypes $AA$, $Aa$, or $aa$. It is important to realize that the genotypes of different individuals (whether in the same or in different generations) will not be independent of each other if they have a common ancestor in any generation after $F_1$. (A common ancestor in $F_1$ has no effect on independence since all $F_1$ individuals are genetically identical.)

Suppose for example that 100 plants are raised in each generation of continued selfing and consider two experimental schemes. In scheme 1 all the plants in $F_k$ are selfed and one seed is raised from each plant to form the next generation. In scheme 2 ten of the plants in $F_k$ are chosen at random and selfed, and ten seeds from each of these plants are raised to form the next generation. Under scheme 1 the genotypes of all plants in the same generation are statistically independent; the genotypes of plants in different generations are only independent if they belong to different lines, that is to say if they can be traced back to different $F_2$ ancestors. Under scheme 2 the genotypes in $F_2$ are statistically independent (since all $F_1$ plants are genetically identical) but in subsequent generations the genotypes of plants in the same generation are only independent if they can be traced back to different $F_2$ ancestors.

We have so far considered only a single locus, but the results are easily extended to any number of unlinked loci since the genotypes at different loci in an individual are statistically independent in all generations after the initial cross; the reason is that the gametic output from $F_1$ contains all possible combinations of genes in equal frequencies, and the independence is maintained afterwards by independent segregation of unlinked loci. The situation becomes more complicated under linkage, and we shall only consider a pair of linked loci with alleles $A$, $a$ at the first locus and $B$, $b$ at the second locus; we suppose that the initial mating is $AABB \times aabb$. In subsequent generations there are nine genotypes which may be phenotypically different (four double homozygotes, four single homozygotes and the double heterozygote), but it is also necessary to distinguish between the coupling and repulsion phases of the double heterozygote ($AB/ab$ and $Ab/aB$) because they have different gametic outputs. All the gametes from $AABB$ individuals are $AB$, and likewise for the other double homozygotes. The gametes from $AaBB$ individuals are half $AB$ and half $aB$, with similar results for the other single homozygotes. The double heterozygote, $AaBb$, produces all four types of gamete, $AB$, $Ab$, $aB$, and $ab$. If the genes are in coupling ($AB/ab$), and if the recombination fraction between the two loci is $r$, the frequencies of these four gametes are $\frac{1}{2}(1-r)$, $\frac{1}{2}r$, $\frac{1}{2}r$ and $\frac{1}{2}(1-r)$ respectively; if the genes are in repulsion ($Ab/aB$) the frequencies are $\frac{1}{2}r$, $\frac{1}{2}(1-r)$, $\frac{1}{2}(1-r)$, and $\frac{1}{2}r$.

Under selfing it is convenient to group the ten genotypes into five classes: (1) $AABB$ or $aabb$, (2) $AAbb$ or $aaBB$, (3) $AABb$ or $aaBb$ or $AaBB$ or $Aabb$, (4) $AB/ab$, and (5) $Ab/aB$. The matrix of transition probabilities is

$$\mathbf{T} = \begin{bmatrix} 1 & 0 & \frac{1}{4} & \frac{1}{2}(1-r)^2 & \frac{1}{2}r^2 \\ 0 & 1 & \frac{1}{4} & \frac{1}{2}r^2 & \frac{1}{2}(1-r)^2 \\ 0 & 0 & \frac{1}{2} & 2r(1-r) & 2r(1-r) \\ 0 & 0 & 0 & \frac{1}{2}(1-r)^2 & \frac{1}{2}r^2 \\ 0 & 0 & 0 & \frac{1}{2}r^2 & \frac{1}{2}(1-r)^2 \end{bmatrix} \tag{3.8}$$

These probabilities are easily obtained by considering the gametic outputs of the genotypes. From symmetry, the frequencies of the genotypes within a class are equal.

Classes 1 and 2 are absorbing states, 3, 4, and 5 are transient states. To find the chance of ultimate absorption in the absorbing states, we write $\mathbf{T}$ as a partitioned matrix

$$\mathbf{T} = \begin{bmatrix} \mathbf{I} & \mathbf{R} \\ \mathbf{O} & \mathbf{Q} \end{bmatrix} \tag{3.9}$$

where $\mathbf{Q}$ is $3 \times 3$ and $\mathbf{R}$ is $2 \times 3$. The chance of ultimate absorption in class $i$ ($i = 1, 2$) starting in class $j + 2$ ($j = 1, 2, 3$) is given by the $(i, j)$th element in the matrix

$$\mathbf{R}(\mathbf{I} - \mathbf{Q})^{-1}. \tag{3.10}$$

(See Kemeny and Snell 1960, Chapter 3.) From this formula we find, after a little algebra, that the probabilities of absorption in classes 1 and 2, starting in class 4 in $F_1$, are $1/(1 + 2r)$ and $2r/(1 + 2r)$ respectively.

The latent roots of the transition matrix are $1$, $\frac{1}{2}$, $(\frac{1}{2} - r)$, and $[\frac{1}{2} - r(1 - r)]$. 1 is a double root, the rest single roots. If $q_i(k)$ is the frequency of the $i$th genotype class in $F_k$, we may write

$$q_i(k) = a_i + b_i(\tfrac{1}{2})^k + c_i(\tfrac{1}{2} - r)^k + d_i[\tfrac{1}{2} - r(1 - r)]^k. \tag{3.11}$$

(The general rule is that if $\lambda$ is a latent root with multiplicity $m$, the coefficient of $\lambda^k$ is a polynomial in $k$ of degree $m - 1$; the first term in (3.11) should be $a_i + a_i^* k$ but $a_i^*$ must be zero since the frequencies must remain bounded.) The constant $a_i$ is the chance of ultimate absorption in class $i$, which has already been found. The remaining constants can be evaluated by calculating the frequencies directly in $F_1$, $F_2$, and $F_3$, thus obtaining three linear equations in three unknowns. The results are given in Table 3.2.

The genotype frequencies at two linked loci in backcross and related generations can be calculated from the gametic outputs of the parents. Consider for example the backcross $B_1 = F_1 \times P_1$. $F_1$ individuals are of

TABLE 3.2

*Genotype frequencies at two linked loci under selfing*

| Genotype | Coefficient of ... | | | |
|---|---|---|---|---|
| | $1$ | $(\tfrac{1}{2})^k$ | $(\tfrac{1}{2} - r)^k$ | $[\tfrac{1}{2} - r(1 - r)]^{k-1}$ |
| AABB, aabb | $\frac{1}{2}(1 + 2r)^{-1}$ | $-1$ | $-\frac{1}{2}(1 + 2r)^{-1}$ | $\frac{1}{4}$ |
| AAbb, aaBB | $r(1 + 2r)^{-1}$ | $-1$ | $\frac{1}{2}(1 + 2r)^{-1}$ | $\frac{1}{4}$ |
| AABb, AaBB, Aabb, aaBb | $0$ | $1$ | $0$ | $-\frac{1}{2}$ |
| AB/ab | $0$ | $0$ | $(1 - 2r)^{-1}$ | $\frac{1}{2}$ |
| Ab/aB | $0$ | $0$ | $-(1 - 2r)^{-1}$ | $\frac{1}{2}$ |

genotype $AB/ab$, and they produce all four types of gamete, $AB$, $Ab$, $aB$, and $ab$, with frequencies $\frac{1}{2}(1-r)$, $\frac{1}{2}r$, $\frac{1}{2}r$, and $\frac{1}{2}(1-r)$ respectively. $P_1$ individuals produce only $AB$ gametes. Thus there will be four genotypes in $B_1$, $AABB$, $AABb$, $AaBB$, and $AaBb$ (in coupling phase, $AB/ab$), with frequencies $\frac{1}{2}(1-r)$, $\frac{1}{2}r$, $\frac{1}{2}r$, and $\frac{1}{2}(1-r)$ respectively. The genotype frequencies in other backcross and related generations are shown in Table 3.3. Because of the complexity of the problem we shall not consider the genotype frequencies under sib-mating in the presence of linkage.

TABLE 3.3

*Genotype frequencies at two linked loci. All frequencies should be divided by 8; r is the recombination fraction,* $s = (1-r)$, $t = (1-2r)$

| Generation | | | | Genotype | | | |
|---|---|---|---|---|---|---|---|
| | $AABB$ | $aabb$ | $AAbb$ $aaBB$ | $AABb$ $AaBB$ | $Aabb$ $aaBb$ | $AB/ab$ | $Ab/aB$ |
| $F_2$ | $2s^2$ | $2s^2$ | $2r^2$ | $4rs$ | $4rs$ | $4s^2$ | $4r^2$ |
| $F_3$ | $2s+r^4+s^4$ | $2s+r^4+s^4$ | $2r(1+rs)$ | $4rs(1-rs)$ | $4rs(1-rs)$ | $2(r^4+s^4)$ | $4r^2s^2$ |
| $B_1$ | $4s$ | — | — | $4r$ | — | $4s$ | — |
| $B_1 \times P_1$ | $4+2s^2$ | — | — | $2r(1+s)$ | — | $2s^2$ | — |
| $B_1 \times P_2$ | — | $2s^2$ | — | — | $2r(1+s)$ | $4+2s^2$ | — |
| $B_1$ selfed | $2+2s+s^3$ | $s^3$ | $r(1+rs)$ | $2r(1+s^2)$ | $2rs^2$ | $2s^3$ | $2r^2s$ |
| $F_2 \times P_1$ | $2(1+st)$ | — | — | $2(1-st)$ | — | $2(1+st)$ | — |
| $F_2 \times F_1$ | $s(1+st)$ | $s(1+st)$ | $r(1-st)$ | $1-st^2$ | $1-st^2$ | $2s(1+st)$ | $2r(1-st)$ |

**Outbred populations under random mating**

In this section we shall consider an outbreeding population in which individuals mate at random with respect to their genotypes. We assume that the population size is effectively infinite so that stochastic fluctuations can be ignored and that there is no mutation or selection. Under these assumptions the structure of genotype frequencies in the population is embodied in the laws of Hardy–Weinberg and linkage equilibrium which will now be described. The effects of departures from these assumptions will be discussed in later chapters.

*Hardy–Weinberg equilibrium*

Consider an autosomal locus with an arbitrary number of alleles, $B_1$, $B_2$, $B_3$, …. The genotype of a diploid individual will be an expression like $B_1B_3$. In this representation it is convenient to assume that the individual has received the first allele from his father and the second from his mother, so that $B_1B_3$ and $B_3B_1$ are to be regarded as distinct, though they will be phenotypically the same; such a genotype will be called an ordered genotype. Write $p_i$ for the relative frequency of the allele $B_i$ in the population. In the absence of selection, migration, mutation and genetic drift, this frequency will remain constant from one generation to the next since these are the only forces which can change it.

The Hardy–Weinberg law states that the probability of the ordered genotype $B_i B_j$ under random mating is

$$Pr(B_i B_j) = p_i p_j. \tag{3.12}$$

With discrete, non-overlapping generations the equilibrium (3.12) is achieved immediately after one generation of random mating, starting from arbitrary genotype frequencies, provided only that the gene frequencies in the two sexes (in a species with separate sexes) are the same. This follows immediately from the observation that random mating of individuals is genetically equivalent to random union of gametes; we may thus imagine that the parents produce a large pool of male gametes and a large pool of female gametes and that the off-spring are obtained by choosing a gamete from each pool at random. The two genes at any locus in the offspring generation must therefore be statistically independent of each other. If the initial gene frequencies in the two sexes are different, say $p_i^*$ in males and $p_i^{**}$ in females, then in the next generation

$$Pr(B_i B_j) = p_i^* p_j^{**}$$

in both sexes. Hardy–Weinberg equilibrium (3.12) will be attained in the following generation with $p_i = \frac{1}{2} p_i^* + \frac{1}{2} p_i^{**}$.

To illustrate the Hardy–Weinberg law, Table 3.4 shows the distribution of phosphoglucomutase types in 537 human families (Harris 1975, p. 62). Phosphoglucomutase is an enzyme which exists in two forms, 1 and 2, which can be distinguished by electrophoresis (a method of separating substances

TABLE 3.4a

*The distribution of phosphoglucomutase types in 537 families*

| Parents | Number of matings | | Children | | |
|---------|---------|-----------|----|----|----|
| | Observed | Predicted† | 11 | 12 | 22 |
| 11 × 11 | 199 | 188 | 392 | — | — |
| 11 × 12 | 203 | 223 | 207 | 215 | — |
| 11 × 22 | 34 | 36 | — | 71 | — |
| 12 × 12 | 77 | 67 | 35 | 81 | 41 |
| 12 × 22 | 21 | 21 | — | 24 | 31 |
| 22 × 22 | 3 | 2 | — | — | 13 |

† Assuming random mating given the observed frequencies in Table 3.4b.

TABLE 3.4b

*The distribution of phosphoglucomutase types in the 1074 parents*

| Type | 11 | 12 | 22 |
|------|----|----|----|
| Observed frequency | 635 | 378 | 61 |
| Expected frequency (assuming Hardy–Weinberg equilibrium) | 632 | 384 | 58 |

with different electric charge). There are three types of individuals, those who have only the first form (type 11), those who have both forms (type 12), and those who have only the second form (type 22). The genetic hypothesis is that the two forms of enzyme are produced by two allelic genes, types 11 and 22 being the two homozygotes and type 12 the heterozygote. The familial data in Table 3.4a are in excellent agreement with this hypothesis. The frequencies of different mating types among the parents in Table 3.4a are in agreement with the frequencies predicted on the assumption that parents mate at random with respect to this locus ($\chi^2 = 4.5$ with three degrees of freedom), and it can be concluded that there is no tendency for assortative mating to occur.

We would therefore expect the genotype frequencies in the population to be in Hardy–Weinberg equilibrium. Table 3.4b shows that there is excellent agreement between the observed genotype frequencies in the 1074 parents and the frequencies predicted by the Hardy–Weinberg law ($\chi^2 = 0.3$ with one degree of freedom). To calculate the Expected frequencies in this table, the first step was to estimate the frequencies of the two alleles, $p_1$ and $p_2$. This was a simple matter of counting genes, since the 1074 diploid individuals have 2148 genes, of which $2 \times 635 + 378 = 1648$ are allele 1 and the remainder are allele 2; hence $p_1 = 0.767$ and $p_2 = 0.233$. We now predict the relative frequencies of the types 11, 12, and 22, assuming Hardy–Weinberg equilibrium, as $p_1^2$, $2p_1p_2$ and $p_2^2$ respectively. Observe that, as in most practical applications, the genotypes in Table 3.4 are unordered; the predicted frequency of the unordered genotype 12 is $2p_1p_2$ since it comprises the ordered genotypes 12 and 21.

So far we have considered an autosomal locus. Sex-linked loci, situated on the X chromosome, differ because males have only one X chromosome whereas females have two. Consider a sex-linked locus with alleles $B_1, B_2, \ldots$. The genotype of a male is a single allele $B_i$, whereas the genotype of a female is an ordered pair, $B_iB_j$. Write $p_i$ for the relative frequency of the allele $B_i$ among all genes at the locus in both sexes. The analogue of the Hardy–Weinberg law states that under random mating the frequency of the genotype $B_i$ in males will be $p_i$ while the frequency of the ordered genotype $B_iB_j$ in females will be $p_ip_j$. This equilibrium will be attained immediately after one generation of random mating if the initial gene frequencies in the two sexes are the same. If they are different, however, the situation is more complicated.

Write $p_i^*(t)$ and $p_i^{**}(t)$ for the frequencies of $B_i$ in males and females in generation $t$. The genotype frequencies in females can be expressed in terms of the gene frequencies of the previous generation:

$$Pr(B_iB_j, t) = p_i^*(t-1)p_j^{**}(t-1).$$

In males, genotype and gene frequencies are equivalent. The recurrence relationships for the gene frequencies are

$$p_i^*(t+1) = p_i^{**}(t)$$
$$p_i^{**}(t+1) = \tfrac{1}{2}p_i^*(t) + \tfrac{1}{2}p_i^{**}(t) \qquad (3.13)$$

since males receive their X chromosomes from their mothers, whereas females receive an X chromosome from each of their parents. Since females have twice as many X chromosomes as males, the average gene frequency over both sexes is defined as

$$p_i(t) = \tfrac{1}{3}p_i^*(t) + \tfrac{2}{3}p_i^{**}(t).$$

It is easily verified that this quantity remains constant from one generation to the next, and may therefore simply be written as $p_i$. The solution of eqn (3.13) is

$$p_i^*(t) = p_i + \tfrac{2}{3}(-\tfrac{1}{2})^t d_i$$
$$p_i^{**}(t) = p_i - \tfrac{1}{3}(-\tfrac{1}{2})^t d_i \qquad (3.14)$$

where

$$d_i = p_i^*(0) - p_i^{**}(0).$$

Thus the gene frequencies in the two sexes do not assume a common value in one generation, as they do at an autosomal locus, but oscillate about their eventual common value with diminishing amplitude. In fact the difference between the gene frequencies, $p_i^*(t) - p_i^{**}(t)$, is halved in magnitude and also changes sign in each generation; this fact can be shown by taking the difference of the two equations in (3.13).

## Linkage equilibrium

We now consider a pair of autosomal loci with alleles $B_1, B_2, B_3 \ldots$ at the first locus and $C_1, C_2, C_3 \ldots$ at the second locus. The genotype will be an expression like $B_1 C_2/B_3 C_1$. In this representation $B_1$ and $C_2$ are derived from the individual's father, $B_3$ and $C_1$ from his mother. $B_1 C_2$ is called the paternal haplotype (*haploid genotype*); $B_3 C_1$ is the maternal haplotype. If the loci are linked, genes in the same haplotype are on the same chromosome, while genes in opposite haplotypes are on homologous chromosomes. If the loci are not linked, genes in the same haplotype are on non-homologous chromosomes inherited in the same gamete.

By the random union of gametes argument it follows that the two haplotypes will become statistically independent of each other after one generation of random mating, starting from arbitrary genotype frequencies. In other words, after one generation any genotype frequency is the product of the corresponding haplotype frequencies. (If the initial genotype frequencies are different in the two sexes, the paternal and maternal haplotype frequencies will be different in the first generation but will become the same in the second generation of random mating.) Thus statistical independence between genes or groups of genes in opposite haplotypes, derived from different parents, is achieved immediately after one generation of random mating; this can be regarded as a generalization of the Hardy–Weinberg law.

By contrast, statistical independence between genes in the same haplotype, inherited from the same parent, is only achieved gradually after several generations of random mating. This type of independence is known under the rather misleading name of *linkage equilibrium*. Write $p_i$ and $q_j$ for the frequencies of the alleles $B_i$ and $C_j$, and write $P_{ij}(t)$ for the frequency of the haplotype $B_i B_j$ in the $t$th generation of random mating. We already know that the frequency of the alleles $B_i$ and $C_j$ in opposite haplotypes is $p_i q_j$ for $t \geqslant \tau$, where $\tau = 1$ if the initial genotype frequencies at $t = 0$ are the same in both sexes and $\tau = 2$ if they are different. If $r$ is the recombination fraction between the two loci it is clear that

$$P_{ij}(t + 1) = (1 - r)\, P_{ij}(t) + rp_i\, q_j, \tag{3.15}$$

for $t \geqslant \tau$, since the haplotype frequencies in generation $t + 1$ are the frequencies of the corresponding gametes produced by the individuals of generation $t$ after recombination. The solution of this recurrence relation is

$$P_{ij}(t) = p_i q_j + (1 - r)^{t-\tau} [P_{ij}(\tau) - p_i q_j]. \tag{3.16}$$

At equilibrium $P_{ij} = p_i q_j$, so that the two genes are statistically independent as expected. However, this equilibrium is only achieved gradually; the departure from equilibrium declines geometrically at a rate equal to the recombination fraction. Genes on different chromosomes behave as if $r = \frac{1}{2}$, so that in this case the departure from equilibrium is halved in each generation.

It was assumed in the last paragraph that the recombination fraction is the same in both sexes. This assumption may not be true; in particular, there is no recombination between genes on the same chromosome in male *Drosophila*. Let the recombination fractions in males and females be $r^*$ and $r^{**}$ respectively; write $P_{ij}^*$ and $P_{ij}^{**}$ for the frequencies of the paternal and maternal haplotypes $B_i C_j$; and write $r$ and $P_{ij}$, without asterisks, for the average values of these quantities, so that $r = \frac{1}{2}(r^* + r^{**})$ and likewise for $P_{ij}$. Since the genotype frequencies in the two sexes are the same, it is clear that

$$P_{ij}^*(t + 1) = (1 - r^*)\, P_{ij}(t) + r^* p_i\, q_j$$
$$P_{ij}^{**}(t + 1) = (1 - r^{**})\, P_{ij}(t) + r^{**} p_i\, q_j.$$

Adding these two equations, we find that eqns (3.15) and (3.16) remain valid, with $r$ and $P_{ij}$ interpreted as average values. Thus the average departure from equilibrium declines geometrically at a rate equal to the average of the recombination fractions.

The above arguments can be extended to any number of loci. Statistical independence between opposite haplotypes is achieved immediately after one generation of random mating by the random union of gametes argument; thus any genotype frequency is equal to the product of the corresponding haplotype frequencies. (If the initial genotype frequencies are different in the two sexes, the paternal and maternal haplotype frequencies will be different in the first generation of random mating; they will become the same in the second generation for autosomal loci but only gradually for sex-linked loci.) Statistical independence between loci in the same haplotype (linkage equilibrium) is achieved gradually after several generations of random mating, though the details become rather complicated for three or more loci (Crow and Kimura 1970, Chapter 2).

## Identity by descent

An important tool in the study of genotype frequencies is the concept of identity by descent developed by Cotterman (1940) and Malécot (1948). Two genes are said to be identical by descent (or simply identical) if one of them has been derived by direct replication from the other or if they are both copies of the same gene in a common ancestor. Identical genes must, barring mutation, be alike in state, that is to say they must be the same alleles. It is this fact which causes the resemblance between relatives, who share some pairs of identical genes; it also causes homozygosity in inbred individuals who are the offspring of related parents and who may in consequence have two identical genes, one inherited from each parent, at the same locus. In this section we shall discuss inbreeding and the resemblance between relatives from the point of view of identity by descent.

### The coefficient of inbreeding

The coefficient of inbreeding of an individual, denoted by $f$, is defined to be the probability that two genes at the same locus are identical by descent. This probability can be calculated by the usual rules of probability from an individual's pedigree. We shall now compute the coefficient of inbreeding under selfing and under brother–sister mating.

We first consider continued selfing. Write $f_n$ for the coefficient of inbreeding in the $n$th generation of selfing, defined as the probability that two genes at the same locus in an individual chosen at random in that generation are identical by descent. $f_n$ obeys the recurrence relationship

$$f_{n+1} = \tfrac{1}{2} + \tfrac{1}{2}f_n.  \tag{3.17}$$

To establish this relationship we observe that the two genes in an individual in generation $n + 1$ may be derived either from the same parental gene or from two different genes in the same individual in generation $n$. These events are equally likely; in the first case the two genes are certainly identical by descent, and in the second case the probability that they are identical is $f_n$. The solution of eqn (3.17) with the initial condition $f_0 = 0$ is

$$f_n = 1 - (\tfrac{1}{2})^n.  \tag{3.18}$$

We now turn to brother–sister mating. The recurrence relationship for the coefficient of inbreeding in this situation is

$$f_{n+2} = \tfrac{1}{4} + \tfrac{1}{2}f_{n+1} + \tfrac{1}{4}f_n. \tag{3.19}$$

To show this we observe that identity between two genes in an individual in generation $n + 2$ can arise in three different ways (see Fig. 3.1):

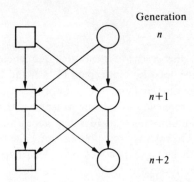

Generation

$n$

$n+1$

$n+2$

Fig. 3.1. A diagram of brother–sister mating; males are square, females are round.

(i) With probability $\tfrac{1}{4}$ they are copies of the same gene in a grandparent in generation $n$; in this case they are certainly identical.

(ii) With probability $\tfrac{1}{4}$ they are derived from two different genes in the same grandparent in generation $n$; in this case the probability that they are identical is $f_n$.

(iii) With probability $\tfrac{1}{2}$ they are derived from different grandparents in generation $n$; in this case the probability that they are identical is $f_{n+1}$ since the two grandparents were mated to produce the individual's parents in generation $n + 1$. (Under brother–sister mating an individual only has two grandparents.)

To obtain the appropriate initial conditions we suppose that there is no inbreeding in generation 0, and that brother–sister mating starts in this generation, the first inbred individuals appearing in generation 1; under this scheme the initial conditions are $f_0 = 0, f_1 = \tfrac{1}{4}$. The roots of the homogeneous equation are 0.809 and −0.309, and the solution of eqn (3.19) with these initial conditions is

$$f_n = 1 - 0.947(0.809)^n - 0.053(-0.309)^n. \tag{3.20}$$

The inbreeding coefficient can be used to find genotype frequencies. Consider an autosomal locus with alleles $B_1, B_2, \ldots$, with frequencies $p_1, p_2, \ldots$. Suppose that $f_0 = 0$ and that the population is in Hardy–Weinberg equilibrium in generation 0 so that the frequency of the ordered genotype $B_i B_j$ is $p_i p_j$.

In a subsequent generation with inbreeding coefficient $f$, the genotype probabilities are

$$Pr(B_i B_j) = (1 - f) p_i p_j, \quad i \neq j$$

$$Pr(B_i B_i) = f p_i + (1 - f) p_i^2. \tag{3.21}$$

The argument is that the probability of observing the allele $B_i$ remains constant from one generation to the next in the absence of directional forces such as selection and migration; and that the two genes must be alike in state if they are identical, whereas they occur independently of each other if they are not identical by descent.

We can use this result to find the genotype frequencies under selfing and brother–sister mating following a cross between two pure lines. The genotype frequencies are in Hardy–Weinberg equilibrium with equal gene frequencies in $F_2$ (or $S_2$), so that we take $f = 0$ in this generation. (It should be realized that identity by descent is always measured with respect to an initial base population in which none of the genes are identical; we may imagine that the genes in this generation are all assigned different labels, and that genes with the same label in subsequent generations are identical by descent. We are free to choose this initial population according to convenience.) Since we are wiping the slate clean in $F_2(S_2)$ we must ensure that the system of inbreeding does not depend on any genetic relations prior to this generation. Thus under sib-mating we suppose that individuals are mated at random in $S_2$, and that sib-mating does not start until $S_3$; as remarked on page 29, random mating in $S_2$ is equivalent to sib-mating since all $F_1$ individuals are heterozygous. Thus we take $F_2$ as generation 0 under selfing and $S_3$ as generation 0 under sib-mating. By using eqn (3.21) in conjunction with eqns (3.18) and (3.20) with $n = k - 2$ under selfing and $n = k - 3$ under sib-mating, we reproduce the results (3.2) and (3.7). The derivation of the genotype frequencies under sib-mating by this method is less direct than the method used on page 31, but avoids the spectral resolution of the transition matrix of the mating type frequencies.

## The relationship between relatives

The concept of identity by descent also provides a useful method of quantifying the genetical relationship between relatives which underlies the observable correlation between them. The results will be needed in analysing the correlations between relatives in an outbred population in Chapter 6. It will be assumed in the discussion that inbreeding does not occur; to justify this assumption we presuppose a large outbred population in which there is a negligible chance that relatives will mate.

Consider a pair of related individuals, such as father and child or a pair of cousins. The seven possible relationships between their ordered genotypes at an autosomal locus with respect to identity by descent are shown in Fig. 3.2, in which identical genes are linked by a line; horizontal lines are not allowed since

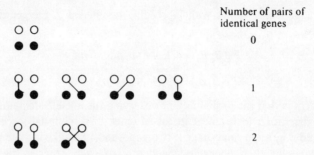

Fig. 3.2. Possible relationships between the ordered genotypes of two non-inbred relatives at a single locus. The genes of one individual are represented by open circles and those of the other by closed circles. Identical genes are linked by a line.

there is no inbreeding. For most purposes the order of the genotype is irrelevant; if it is ignored there are only three different relationships depending on whether there are 0, 1, or 2 pairs of identical genes.

It is usually impossible to predict exactly how many pairs of identical genes a pair of relatives will share; the best that can be done is to evaluate the probabilities that there will be 0, 1, or 2 pairs of identical genes at a given locus. These probabilities will be denoted $P_0$, $P_1$, and $P_2$ respectively after a slight modification of Cotterman's notation. They are tabulated in Table 3.5 for some common relatives.

A mother transmits one gene at any locus to each of her children; thus mother and child are certain to have exactly one pair of identical genes at every locus. The child is equally likely to transmit a maternal gene (identical to one of its mother's genes) or an unrelated paternal gene to its children, so that grandmother and grandchild are equally likely to have one pair of identical genes or none. A pair of sibs is equally likely to have identical or non-identical maternal genes; likewise they are equally likely to have identical or non-identical paternal genes. Since these events are independent, sibs will have 0, 1, or 2 pairs of identical genes with probabilities $\frac{1}{4}$, $\frac{1}{2}$, and $\frac{1}{4}$. By the same argument, half-sibs (which have only one parent in common) are equally likely to have 0 or 1 pair of identical genes. The other results in Table 3.5 can be verified in a similar way.

TABLE 3.5

*The distribution of pairs of identical genes in some common relatives*

| Relationship | $P_0$ | $P_1$ | $P_2$ |
|---|---|---|---|
| Identical twins | 0 | 0 | 1 |
| Parent–child | 0 | 1 | 0 |
| Sibs | 0.25 | 0.5 | 0.25 |
| Grandparent–grandchild, half-sibs, uncle–nephew | 0.5 | 0.5 | 0 |
| Greatgrandparent–greatgrandchild, cousins | 0.75 | 0.25 | 0 |

It is sometimes necessary to consider the simultaneous occurrence of identity by descent at two or more loci. Under random mating the numbers of pairs of identical genes at loci on different chromosomes will be independent because of independent assortment of chromosomes, but this will no longer be true for linked loci. Consider a pair of sibs as an example. Suppose that the mother has the genotype $B_1 C_1/B_2 C_2$ at a pair of linked loci with recombination fraction $r$, where the different subscripts on $B_1$ and $B_2$ indicate that they are non-identical, though they may be alike in state. She will produce four types of gametes, $B_1 C_1$, $B_1 C_2$, $B_2 C_1$, and $B_2 C_2$ with frequencies $\frac{1}{2}(1 - r)$, $\frac{1}{2}r$, $\frac{1}{2}r$, and $\frac{1}{2}(1 - r)$ respectively; these are also the maternal haplotype frequencies of her children. Hence we can calculate the probabilities that the maternal haplotypes of a pair of her children will be identical at both loci, at $B$ but not $C$, at $C$ but not $B$, and at neither locus. These probabilities are $\frac{1}{2} - r(1 - r)$, $r(1 - r)$, $r(1 - r)$, and $\frac{1}{2} - r(1 - r)$ respectively; as expected there is lack of independence unless $r = \frac{1}{2}$. The same result holds independently for the paternal haplotype, whence the joint distribution of the numbers of pairs of identical genes at the two loci in sibs can be calculated. For example, the probability that there are no identical genes at either locus is $(\frac{1}{2} - r(1 - r))^2$, which is greater than $\frac{1}{16}$, the value under independence, unless $r = \frac{1}{2}$. It should be noted that parent and child are an exception to the general rule that the numbers of pairs of identical genes at linked loci are not independent since they always have exactly one pair of identical genes at every locus.

These concepts can be extended to sex-linked loci, though the sexes of the relatives and of any intervening individuals in the pedigree must be taken into account. Since sex-linked genes are diploid in females and haploid in males, a pair of female relatives may share 0, 1, or 2 pairs of identical genes at a sex-

TABLE 3.6

*The distribution of pairs of identical genes at a sex-linked locus in some common relatives*

| Relationship | $P_0$ | $P_1$ | $P_2$ |
|---|---|---|---|
| Mother–daughter | 0 | 1 | 0 |
| Sisters | 0 | 0.5 | 0.5 |
| Paternal halfsisters | 0 | 1 | 0 |
| Maternal halfsisters | 0.5 | 0.5 | 0 |
| Father–son | 1 | 0 | |
| Brothers | 0.5 | 0.5 | |
| Paternal halfbrothers | 1 | 0 | |
| Maternal halfbrothers | 0.5 | 0.5 | |
| Mother–son | 0 | 1 | |
| Father–daughter | 0 | 1 | |
| Brother–sister | 0.5 | 0.5 | |
| Paternal halfbrother–halfsister | 1 | 0 | |
| Maternal halfbrother–halfsister | 0.5 | 0.5 | |

linked locus, as at an autosomal locus, but male relatives or a male–female pair can have at most one pair of identical genes. The probability distributions for parent and child and for full and half sibs are shown in Table 3.6. (Note that calculating the number of pairs of identical genes ignores the ordering of the genotype and is only appropriate if the maternal or paternal origin of the X chromosome in a female is biologically irrelevant. These calculations are not appropriate to kangaroos and other marsupials in which the paternal X chromosome is always inactivated.)

## Problems

1. Following a cross between two pure lines, an $F_3$ plant is crossed to an unrelated $F_2$ plant, and two plants are raised from the cross. What are the chances that (a) one of these plants (chosen at random) is heterozygous at a specified locus, (b) they are both heterozygous? What do these chances become if an $F_3$ plant is crossed back to its $F_2$ parent?

2. Evaluate the constants $b_i$, $c_i$, $d_i$ in eqn (3.6) and hence verify eqn (3.7).

3. Find from first principles the genotype frequencies at two linked loci in the generation resulting from the cross $F_2 \times F_1$.

4. Acid phosphatase is an enzyme which exists in man in three forms, $A$, $B$, and $C$, which can be distinguished by electrophoresis. The frequencies of the six possible phenotypes in 880 individuals are shown below:

| Phenotype | $AA$ | $BB$ | $CC$ | $AB$ | $AC$ | $BC$ |
|-----------|------|------|------|------|------|------|
| Frequency | 119 | 282 | 0 | 379 | 39 | 61 |

Estimate the gene frequencies and calculate the expected phenotype frequencies on the assumption that the population is in Hardy–Weinberg equilibrium. (See Harris 1975, p. 182.)

5. In an experimental *Drosophila* population a number of white-eyed males are mated with normal, wild type females. Find the frequencies of white-eyed males and females for six generations. (White eye is a sex-linked recessive character.)

6. A population of mice is homozygous for the alleles $B$ and $C$ at two autosomal loci, while another population is homozygous for the alternative alleles $b$ and $c$. Suppose that these two populations are of equal size, and that they meet and mate at random. Find the frequency of the haplotype $BC$ in the first four generations if the recombination fraction is (a) $\frac{1}{2}$ (b) $\frac{1}{4}$. Repeat these calculations if individuals from opposite populations are mated together initially.

7. The coefficient of relationship between two relatives in an outbred population is defined as $R = \frac{1}{2}P_1 + P_2$. If these relatives mate, show that $\frac{1}{2}R$ is the coefficient of inbreeding among their offspring. Hence find the coefficient of inbreeding among the offspring of cousin marriages.

   A recessive condition has a frequency of 1 in 10 000 in the general population, which may be assumed to be in Hardy–Weinberg equilibrium. What is its frequency in the offspring of cousin marriages?

8. Derive the results of Table 3.5. (You may find it helpful to draw the pedigrees first.)

9. Find the joint distribution of pairs of identical genes at two linked loci in sibs; express the results in terms of $a = (1 - 2r)^2$, where $r$ is the recombination fraction.

10. Derive the results of Table 3.6.

# 4 Decomposition of the genotypic value

*The genotypic value G, as defined in Chapter 2 is a function of the genotype, γ. It is only possible to observe the relationship between G(γ) and γ directly for a few special genotypes which can be observed in large numbers (pure lines and F₁ crosses between them). For most genotypes it is necessary to infer this relationship indirectly from statistical analysis of the means and variances in different generations following a line cross or of correlations between relatives in an outbred population. To do this it is convenient to decompose the genotypic value into components representing the effects of additive gene action, dominance, and epistasis; the statistical analysis can be interpreted in terms of these components. The purpose of this chapter is to explain this decomposition.*

## General theory

Consider a character determined by $n$ autosomal loci, each with an arbitrary number of alleles, and write $B_{ij}$ for the $j$th allele at the $i$th locus. A typical ordered genotype may be written in the form

$$\gamma = B_{1j_1} B_{2j_2} \ldots B_{nj_n} / B_{1k_1} B_{2k_2} \ldots B_{nk_n}$$

$$= \text{paternal haplotype/maternal haplotype.} \tag{4.1}$$

If we were able to measure $G(\gamma)$ for all genotypes in this population, we could regard these measurements as a crossed classification (factorial design) with $2n$ factors, using the terminology of the analysis of variance. We could therefore decompose the genotypic value into components representing the main effects of the $2n$ genes and all possible interactions between them. This hypothetical decomposition will serve as a statistical model in terms of which observable means, variances, and correlations can be interpreted. (R. A. Fisher invented the factorial design in 1926 in connection with agricultural experiments, but he later traced the idea to origins in his thought about quantitative inheritance: 'The 'factorial' method of experimentation ... derives its structure, and its name, from the simultaneous inheritance of Mendelian factors.' (Fisher 1952, quoted by Box 1978.))

To formalize this decomposition we first need to establish some notation. Denote by $N$ the set containing the first $2n$ positive integers, $N = \{1, 2, \ldots, 2n\}$. These integers are used as indices to mark the positions of the genes in the genotype, 1 to $n$ denoting the genes in the paternal haplotype and $n + 1$ to $2n$ the genes in the maternal haplotype. If $S$ and $T$ are subsets of $N$, $T \subset S$ means that $T$ is a proper subset of $S$, while $T \subseteq S$ allows the possibility that $T = S$. If

$S$ is any subset of $N$ (including $N$ itself and the null set $\varnothing$), $X_S$ denotes the effect due to the interaction between the genes whose positions are in $S$. If $S$ is the null set, $X_\varnothing$ is the mean value in the reference population; if $S$ contains only one position, say $S = \{r\}$ $X_r$ is the main effect due to the gene in the $r$th position; if $S$ contains two positions, say $S = \{r, s\}$, $X_{rs}$ is the interaction between the genes in those positions; and so on. In general $X_S$ is a function of the genes whose positions are in $S$. The definition of $X_S$ depends on the genotype frequencies in the reference population, but in any case the components sum to $G$ identically if all possible interactions are included. Thus

$$G = \sum_{S \subseteq N} X_S, \tag{4.2}$$

the sum being over all subsets of $N$, including $N$ itself (which gives the highest order interaction) and the null set (which gives the mean value, $m$).

To explain the meaning of this decomposition, consider the case with two loci. The decomposition is

$$G = m + [(X_1 + X_3) + (X_2 + X_4)] + [X_{13} + X_{24}] + [X_{12} + X_{14} + X_{23} + X_{34}]$$
$$+ [(X_{123} + X_{134}) + (X_{124} + X_{234})] + [X_{1234}], \tag{4.3}$$

where similar terms have been grouped together. Consider the five groups in square brackets in turn:

(i) $X_1$ and $X_3$ are the main effects of the two genes at the first locus. Their sum is denoted by $A_1$ and is called the additive component at the first locus. Similarly $A_2 = (X_2 + X_4)$ is the additive component at the second locus.

(ii) $X_{13}$ is the interaction between the main effects at the first locus; it is usually called the dominance deviation at the first locus, denoted by $D_1$. Likewise $D_2 = X_{24}$.

(iii) The four terms in the third bracket all represent interactions between the main effect of a gene at the first locus and the main effect of a gene at the second locus; we denote the sum of these four terms as $AA_{12}$.

(iv) $X_{123}$ and $X_{134}$ both represent interactions between the main effect of a gene at the second locus and the dominance deviation at the first locus; we denote their sum as $AD_{21}$. Likewise we write $AD_{12} = (X_{123} + X_{234})$.

(v) The last term is the interaction between the dominance deviations at the two loci which we denote by $DD_{12}$.

Using this more descriptive notation we can rewrite eqn (4.3) in condensed form as

$$G = m + (A_1 + A_2) + (D_1 + D_2) + AA_{12} + (AD_{12} + AD_{21}) + DD_{12}. \tag{4.4}$$

The terms grouped together in eqn (4.4) represent similar factors arising from different loci, and we can effect a further condensation by adding them together:

$$G = m + A + D + AA + AD + DD, \tag{4.5}$$

where $A$ denotes $(A_1 + A_2)$ and so on. $A$ is the additive genetic value of the genotype, $D$ is the (total) dominance deviation, $AA$ is the additive $\times$ additive interaction, and so on. These ideas can be extended in the obvious way to any number of loci. We define

$$A_i = X_i + X_{i+n}$$

$$D_i = X_{i,i+n}$$

$$AA_{ij} = X_{i,j} + X_{i,j+n} + X_{i+n,j} + X_{i+n,j+n} \qquad (4.6)$$

$$AD_{ij} = X_{i,j,j+n} + X_{i+n,j,j+n}$$

$$DD_{ij} = X_{i,j,i+n,j+n}$$

and so on. Then

$$G = m + \sum_i A_i + \sum_i D_i + \sum_{i<j} AA_{ij} + \sum_{i \neq j} AD_{ij} + \sum_{i<j} DD_{ij}$$

$$+ \sum_{i<j<k} AAA_{ijk} + \cdots \qquad (4.7)$$

$$= m + A + D + AA + AD + DD + AAA + \cdots$$

The above analysis assumes that all the loci are autosomal. If there is sex-linkage the genotypic value must be decomposed separately in the two sexes since females are diploid but males are haploid at sex-linked loci. The decomposition of the female genotypic value proceeds exactly as before, but in the male we must remember that a sex-linked locus is represented only once, in the maternal but not in the paternal haplotype. Consider a sex-linked locus which occupies the $r$th position in the maternal haplotype; we may regard the corresponding position in the paternal haplotype as blank. Then the additive effect at this locus must be defined as the main effect of this gene, $A_r = X_{n+r}$, rather than as the sum of the effects of the paternal and maternal genes, and there is no analogue of the dominance deviation. Similar remarks apply to epistatic interactions involving sex-linked loci in males. (If the male is XX and the female XY as in birds, butterflies and moths, and certain other animals, interchange 'male' with 'female' and 'paternal' with 'maternal' in the above account.) It may also be necessary to decompose the genotypic value separately in the two sexes if the genotype reacts differently in males and females (genotype $\times$ sex interaction).

We now return to the definition of the components in the fundamental decomposition (4.2). $X_S$ is a function of the genes present at the positions in the ordered genotype contained in $S$. Suppose that there are $k_i$ different alleles at the $i$th locus; thus there are $k_r$ possible alleles at the $r$th position in the genotype if we define $k_r = k_{r-n}$ when $r > n$. Denote by $v(S)$ the number of different genotypes generated by different combinations of alleles at positions in $S$, so that

$$v(S) = \prod_{r \in S} k_r.$$

To define the function $X_S$ we need to evaluate the $v(S)$ values of $X_S$ corresponding to different genotypes at positions in $S$.

The number of values to be calculated for all the components is

$$\sum_{S \subseteq N} v(S) = v(N) + \sum_{S \subset N} v(S).$$

But the only information availabe for computing these values is contained in the different genotypic values, whose number is $v(N)$. Thus there is not enough information to calculate all these values. This problem is often met in the analysis of variance when the model contains too many parameters. The solution is to impose a sufficient number of constraints on the components $X_S$ to enable them all to be calculated from the $v(N)$ equations implicit in the identity (4.2).

The constraints which it is appropriate to impose and in consequence the definition of the components $X_S$ depend on the genotype frequencies in the population. In the next section we shall investigate the decomposition appropriate in a population in which there is random mating so that there are no departures from Hardy–Weinberg or linkage equilibrium. This population has rather simple properties because all the $2n$ positions in the genotype are statistically independent of each other; in analysis of variance terminology the design is balanced. On page 54 we derive the decomposition appropriate in an inbred population. It is convenient for brevity to refer to the population of genotype frequencies with reference to which the decomposition is defined as the reference population.

The component $X_S$ is a function of the genes with positions in $S$, and is thus a random variable whose distribution depends on the probabilities of those genes. It may or may not be the case that the actual population of genotype frequencies with which we are dealing is the same as the reference population with respect to which the decomposition was defined. When these two populations are the same the random variable $X_S$ will have particularly simple properties, but it is not always possible to ensure that this is the case. Suppose for example that two pure lines are crossed and then allowed to self-fertilize in successive generations. The genotype frequencies change in each generation, and it is therefore necessary to choose one of them as the reference population for the decomposition, since the parameters in a model must remain constant. It is conventional to choose the limiting state towards which the population tends after many generations of selfing as the reference population for the decomposition (the so-called $F_\infty$ metric), but this choice is to some extent arbitrary. In particular we might choose the $F_2$ generation as the reference population, giving rise to the so-called $F_2$ metric; this is a special case of the random mating decomposition, with two equally frequent alleles at each locus.

## Random mating populations

Let the gene frequency of the allele $B_{ij}$ be $p_{ij}$. Suppose that the population is in Hardy–Weinberg and linkage equilibrium so that the probability of the typical genotype $\gamma$ in eqn (4.1) is

$$Pr(\gamma) = p_{1j_1} p_{2j_2} \cdots p_{nj_n} \cdot p_{1k_1} p_{2k_2} \cdots p_{nk_n}. \qquad (4.8)$$

By analogy with the usual constraints for a balanced crossed classification we impose the constraints

$$\underset{r}{E}(X_S) = 0 \quad \text{if } r \in S \qquad (4.9)$$

where $\underset{r}{E}$ denotes an Expected value with respect to the genotype frequencies (4.8) taken over the possible alleles in the $r$th position.

To evaluate the components $X_S$ under these constraints, we define the quantity

$$G_S = \underset{N-S}{E}(G). \qquad (4.10)$$

$G_S$ is a function of those genes whose positions are in $S$; it is the average genotypic value of individuals with a specified genotype in those positions, averaged over all other positions. We now find from the fundamental decomposition (4.2) that

$$G_S = \underset{N-S}{E} \sum_{T \subseteq N} X_T = \sum_{T \subseteq S} X_T. \qquad (4.11)$$

For if $T \subseteq S$, then $\underset{N-S}{E}(X_T) = X_T$ since $(N - S)$ and $T$ are disjoint. On the other hand, if $T \nsubseteq S$, then there is a number $r$ which belongs to both $T$ and $(N - S)$, so that

$$\underset{N-S}{E}(X_T) = 0 \qquad (4.12)$$

from eqn (4.9); this result assumes that the distribution of $X_T$ is unaffected by knowledge of genes with positions not in $T$, so that it is crucially dependent on the assumption of independence embodied in eqn (4.8). From eqn (4.11) we find the recurrence relationship

$$X_S = G_S - \sum_{T \subset S} X_T.$$

The components $X_S$ can now be defined successively for sets of increasing size, starting with the null set $\varnothing$:

$$
\begin{aligned}
X_\varnothing &= G_\varnothing = m \\
X_r &= G_r - m \\
X_{rs} &= G_{rs} - (m + X_r + X_s),
\end{aligned}
\tag{4.13}
$$

and so on.

The component $X_S$ is a random variable whose distribution is determined by the definition of the function and by the probabilities of the genes with positions in $S$. Under the probabilities (4.8), it follows from eqn (4.9) that $E(X_S) = 0$ if $S$ is not empty. (Note that $E(X_\varnothing) = m$, the mean genotypic value.) If $S$ and $T$ are disjoint then $X_S$ and $X_T$ are independently distributed because of the independence embodied in eqn (4.8). If $S$ and $T$ are not disjoint then $X_S$ and $X_T$ are not necessarily independent, but they are uncorrelated. For if $S \neq T$, we may suppose without loss of generality that there is a number $r$ which belongs to $T$ but not to $S$. Then

$$
\mathrm{Cov}(X_S, X_T) = E(X_S X_T) = \mathop{E}_{N-r} \mathop{E}_{r} (X_S X_T) = \mathop{E}_{N-r} (X_S \mathop{E}_{r} (X_T)) = 0 \tag{4.14}
$$

(If $S$ is not a subset of $T$ this result depends on the independence of genes in $T$ and in $S$ but not in $T$.) Hence the basic decomposition

$$
G = \sum_{S \subseteq N} X_S
$$

provides a corresponding decomposition of the genetic variance

$$
V_G = \sum_{S \subseteq N} \mathrm{Var}(X_S). \tag{4.15}
$$

It is in order to ensure this result that it is convenient to impose the constraints (4.9). From the condensed decomposition (4.7) we may write

$$
\begin{aligned}
V_G &= \sum_i \mathrm{Var}(A_i) + \sum_i \mathrm{Var}(D_i) + \sum_{i<j} \mathrm{Var}(AA_{ij}) \\
&\quad + \sum_{i \neq j} \mathrm{Var}(AD_{ij}) + \cdots \\
&= V_A + V_D + V_{AA} + V_{AD} + V_{DD} + V_{AAA} + \cdots
\end{aligned}
\tag{4.16}
$$

$V_A$ is the additive genetic variance, $V_D$ is the dominance variance, $V_{AA}$ is the additive $\times$ additive variance, and so on. It is these components of variance which can be estimated from correlations between relatives, as we shall see in Chapter 6.

To illustrate these ideas consider a single locus with two equally frequent alleles, $B$ and $b$, with $B$ completely dominant to $b$. The decomposition of the

TABLE 4.1

*Decomposition of the genotypic value at a single locus*

| Ordered genotype | BB | Bb | bB | bb |
|---|---|---|---|---|
| Frequency | 0.25 | 0.25 | 0.25 | 0.25 |
| Genotypic value, $G$ | 1 | 1 | 1 | 0 |
| $G_\emptyset = m$ | 0.75 | 0.75 | 0.75 | 0.75 |
| $G_1$ | 1 | 1 | 0.5 | 0.5 |
| $G_2$ | 1 | 0.5 | 1 | 0.5 |
| $X_1 = G_1 - m$ | 0.25 | 0.25 | −0.25 | −0.25 |
| $X_2 = G_2 - m$ | 0.25 | −0.25 | 0.25 | −0.25 |
| $X_{12} = G - m - X_1 - X_2$ | −0.25 | 0.25 | 0.25 | −0.25 |
| $A = X_1 + X_2$ | 0.5 | 0 | 0 | −0.5 |
| $D = X_{12}$ | −0.25 | 0.25 | 0.25 | −0.25 |

Decomposition of the variance

$\text{Var}(X_1) - \text{Var}(X_2) = \text{Var}(X_{12}) = 0.0625$

$V_A = 0.125$

$V_D = 0.0625$

$V_G = 0.1875$

genotypic value is shown in Table 4.1. It will be seen that $X_1$, $X_2$, and $X_{12}$ are all random variables which take the values $\frac{1}{4}$ and $-\frac{1}{4}$ with equal probabilities. They are uncorrelated and pairwise independent, but they are not mutually independent since $X_{12} = \frac{1}{4} - 2(X_1 + X_2)^2$. $X_1$ and $X_2$ represent the main effects due to the paternal and maternal genes respectively. They are identically and independently distributed, and their sum, $A = (X_1 + X_2)$, represents the contribution to the genotypic value due to additive gene action; its variance $V_A$ is the additive genetic variance. $X_{12} = D$ represents the effect of dominance and its variance $V_D$ is the dominance variance.

This analysis can be extended to the single locus, two allele case with arbitrary gene frequencies and genotypic values. Write $p$ and $q$ for the gene frequencies of $B$ and $b$, and let the genotypic values of $BB$, $Bb$, and $bb$ be $+a$, $d$, and $-a$ respectively; thus $d = a$ corresponds to complete dominance of $B$ and $d = 0$ corresponds to no dominance. It is important to realize that the definition and value of the components in a Table like 4.1 depends on the gene frequency, which determines the genotype frequencies, as well as on the genotypic values. It is not difficult to show that the additive and dominance components of the genetic variance are

$$V_A = 2pq[a + (q - p)d]^2$$
$$V_D = (2\,pqd)^2. \tag{4.17}$$

When there is no dominance $V_D$ is of course zero; the genotypic value is the sum of two independent contributions from the two genes and its variance thus has the binomial form $2\,pqa^2$. When there is complete dominance of $B$, the ratio of the additive to the dominance variance is $V_A/V_D = 2q/p$.

To illustrate the multi-locus case Table 4.2 shows the genotypic values in a

TABLE 4.2

*Decomposition of the genetic variance under a two locus model with complementary gene action*

|  |  | BB | Bb | bB | bb | $G_{24}$ |
|---|---|---|---|---|---|---|
|  | CC | 1 | 1 | 1 | 0 | 0.75 |
|  | Cc | 1 | 1 | 1 | 0 | 0.75 |
| Genotypic values | cC | 1 | 1 | 1 | 0 | 0.75 |
|  | cc | 0 | 0 | 0 | 0 | 0.00 |
|  | $G_{13}$ | 0.75 | 0.75 | 0.75 | 0.00 |  |

Decomposition of the variance

$$V_A = 0.1406$$
$$V_D = 0.0703$$
$$V_{AA} = 0.0156$$
$$V_{AD} = 0.0156$$
$$V_{DD} = 0.0039$$
$$V_G = 0.2461$$

model with two loci each with two alleles with complementary gene action. It will be assumed that the frequency of each allele at each locus is $\frac{1}{2}$. The marginal genotypic values at each locus ($G_{13}$ and $G_{24}$) under this assumption are also shown in the table together with the decomposition of the genetic variance. We shall now describe how this decomposition was obtained.

The marginal genotypic values at each locus show complete dominance of one allele over the other. Adding the additive and dominance variances for both loci we obtain $V_A$ and $V_D$. Turning to the additive × additive component of the interaction, we first compute the function $X_{12}$, which depends on the genes in the paternal haplotype. We find that

$$X_{12}(BC/\text{-}\text{-}) = G_{12}(BC/\text{-}\text{-}) - m - X_1(B\text{-}/\text{-}\text{-}) - X_2(\text{-}C/\text{-}\text{-})$$
$$= 1 - \tfrac{9}{16} - \tfrac{3}{16} - \tfrac{3}{16} = \tfrac{1}{16}$$

and that similarly

$$X_{12}(bc/\text{-}\text{-}) = \tfrac{1}{16}$$
$$X_{12}(Bc/\text{-}\text{-}) = X_{12}(bC/\text{-}\text{-}) = -\tfrac{1}{16}.$$

Thus $X_{12}$ is a random variable taking the values $\frac{1}{16}$ and $-\frac{1}{16}$ with equal frequencies, since the haplotypes are equally likely, and its variance is $\frac{1}{256}$. The three other components of $AA$ ($X_{14}$, $X_{23}$, and $X_{34}$) are identically distributed so that $V_{AA} = \frac{1}{64}$. It can be shown in a similar way that $X_{123}$, $X_{134}$, $X_{124}$, $X_{234}$, and $X_{1234}$ are also equally likely to take the values $\frac{1}{16}$ and $-\frac{1}{16}$; the first four of these variables comprise the additive × dominance interaction and the last one is the dominance × dominance interaction. Thus the genetic variance can be decomposed into the five components shown in Table 4.2.

### Inbred populations

As in the last section we suppose that the gene frequency of the allele $B_{ij}$ is $p_{ij}$, but we shall allow for the possibility that there is partial or complete inbreeding. If the inbreeding coefficient is $f$ the probability of a typical ordered genotype at the $i$th locus can often be written in the form

$$Pr(B_{ij}B_{ik}) = (1 - f)p_{ij}\, p_{ik} + f\,\delta_{jk}\, p_{ij}, \tag{4.18}$$

where $\delta_{jk}$ is the *Kronecker delta function*

$$\delta_{jk} = \begin{cases} 0 & \text{if } j \neq k \\ 1 & \text{if } j = k. \end{cases}$$

We consider a reference population in which genes at different loci are statistically independent.

To find a decomposition appropriate to an inbred population we impose the constraints

$$\underset{r}{E}\,(X_S) = 0 \quad \text{if } r \in S \tag{4.19}$$

as before, but with Expected values taken with respect to the genotype frequencies (4.18). We also define

$$G_S = \underset{N-S}{E}\,(G) \tag{4.20}$$

as before. It can be shown that

$$G_S = \sum_{T \subseteq S} (1 + f)^{c(S,T)}\, X_T \tag{4.21}$$

where $c(S, T)$ is the number of loci which are represented exactly once in both $S$ and $T$. (The $i$th locus is said to be represented once in $S$ if either $i$ or $i + n$, but not both, belong to $S$.) From eqn (4.21) we find a recurrence relationship for $X_S$ as before.

To demonstrate (4.21) in a particular case, consider two loci each with two alleles. We shall verify that $G_{12}$ satisfies (4.21). By definition

$$G_{12} = \sum_{T \subseteq N\,3,4} E\,(X_T | 1, 2).$$

Note that the Expected values are conditional on the genes in positions 1 and 2, whereas the constraints in eqn (4.19) are not conditional on positions outside $S$. We observe that

$$= X_T \quad \text{if } T = \varnothing, 1, 2, \text{ or } 12 \tag{4.22a}$$

$$\underset{3,4}{E}\,(X_T | 1, 2) = 0 \quad \text{if } T = 123, 124, \text{ or } 1234 \tag{4.22b}$$

$$= 0 \quad \text{if } T = 13, 24, 134, \text{ or } 234 \tag{4.22c}$$

Eqn (4.22a) follows from the fact that in these cases $X_T$ does not depend on the genes in positions 3 and 4; eqn (4.22b) follows immediately from eqn (4.19); and eqn (4.22c) follows from eqn (4.19) together with the fact that genes at different loci are statistically independent. We next observe that

$$\underset{3,4}{E}\,(X_3 | 1, 2) = f X_1$$

since $X_3 = X_1$ if the paternal and maternal genes at the first locus are identical by descent, while $X_3$ is independent of $X_1$ and thus has zero Expectation otherwise. Similarly, the conditional Expectations of $X_4$, $X_{14}$, $X_{23}$, and $X_{34}$ are respectively $f X_2, f X_{12}, f X_{12}$, and $f^2 X_{12}$. Hence

$$G_{12} = X_\varnothing + (1 + f)(X_1 + X_2) + (1 + f)^2 X_{12}.$$

When $f = 0$ we obtain the random mating decomposition considered in the last section. When $f = 1$ we obtain the completely inbred decomposition whose components satisfy the equations

$$G_S = \sum_{T \subseteq S} 2^{c(S,T)} X_T. \tag{4.23}$$

This decomposition is illustrated in Table 4.3 for a single locus with two equally frequent alleles with complete dominance; the corresponding random mating decomposition is shown in Table 4.1.

TABLE 4.3

*Decomposition of the genotypic value at a single locus under inbreeding*

| Ordered genotype | BB | Bb | bB | bb |
|---|---|---|---|---|
| Frequency | 0.5 | 0 | 0 | 0.5 |
| Genotypic value, $G$ | 1 | 1 | 1 | 0 |
| $G_\varnothing = m$ | 0.5 | 0.5 | 0.5 | 0.5 |
| $G_1$ | 1 | 1 | 0 | 0 |
| $G_2$ | 1 | 0 | 1 | 0 |
| $X_1 = \frac{1}{2}(G_1 - m)$ | 0.25 | 0.25 | −0.25 | −0.25 |
| $X_2 = \frac{1}{2}(G_2 - m)$ | 0.25 | −0.25 | 0.25 | −0.25 |
| $A = X_1 + X_2$ | 0.5 | 0 | 0 | −0.5 |
| $D = X_{12} = G - m - X_1 - X_2$ | 0 | 0.5 | 0.5 | 0 |

Note: $G_1$ is the value of a homozygote with the paternal gene fixed, while $G_2$ is the value of a homozygote with the maternal gene fixed. $X_1$ and $X_2$ are then found from eqn (4.23).

We shall now consider the reduced decomposition

$$G = m + \sum_i A_i + \sum_i D_i + \sum_{i<j} AA_{ij} + \cdots$$

under the inbred model. The components can be calculated as follows:

$$A_i(B_{ij}B_{ij}) = G_{i,i+n}[B_{ij}B_{ij}] - m$$

$$A_i(B_{ij}B_{ik}) = \tfrac{1}{2}[A_i(B_{ij}B_{ij}) + A_i(B_{ik}B_{ik})], \quad j \neq k$$

$$D_i(B_{ij}B_{ik}) = G_{i,i+n}(B_{ij}B_{ik}) - m - A_i(B_{ij}B_{ik}) \tag{4.24}$$

$$AA_{ij}(B_{ik}B_{jm}B_{jm}) = G_{i,i+n,j,j+n}(B_{ik}B_{ik}B_{jm}B_{im})$$

$$- m - A_i(B_{ik}B_{ik}) - A_j(B_{jm}B_{jm})$$

$$AA_{ij}(B_{ik}B_{il}B_{jm}B_{jp}) = \tfrac{1}{4}[AA_{ij}(B_{ik}B_{ik}B_{jm}B_{jm})$$

$$+ \cdots + AA_{ij}(B_{il}B_{il}B_{jp}B_{jp})]$$

and so on. Note that $D_i = 0$ for any homozygote at the $i$th locus; likewise $AD_{ij} = 0$ for any homozygote at the $j$th locus, and so on.

An important special case of the completely inbred decomposition occurs when there are only two alleles per locus (say $B_i$ and $b_i$ at the $i$th locus) each with a frequency of $\tfrac{1}{2}$. We define the constants

$$a_i = A_i(B_i B_i) = G_{i,i+n}(B_i B_i) - m$$

$$d_i = D_i(B_i b_i) = G_{i,i+n}(B_i b_i) - m$$

$$aa_{ij} = AA_{ij}(B_i B_i B_j B_j) = G_{i,i+n,j,j+n}(B_i B_i B_j B_j) - m - a_i - a_j \tag{4.25}$$

$$ad_{ij} = AD_{ij}(B_i B_i B_j B_j) = G_{i,i+n,j,j+n}(B_i B_i B_j B_j) - m - a_i - a_j$$

and so on. We denote by $z_i$ the number of $B_i$ genes at the $i$th locus ($z_i = 0$, 1, or 2); the unordered genotype can be represented by the vector $\mathbf{z}$. It is now quite easy to show that the components for an arbitrary genotype are

$$A_i = (z_i - 1)a_i$$
$$D_i = z_i(2 - z_i)d_i$$
$$AA_{ij} = (z_i - 1)(z_j - 1)aa_{ij} \tag{4.26}$$
$$AD_{ij} = (z_i - 1)z_j(2 - z_j)ad_{ij},$$

and so on. The factor $(z_i - 1)$ for additive components arises from the fact that $A_i(b_i b_i) = -A_i(B_i B_i)$ which follows from the symmetry of the situation; a similar result holds for interactions involving additive components. The factor $z_j(2 - z_j)$ for dominance components and their interactions expresses the fact that these components are zero when the corresponding loci are homozygous.

The constants $a_i$ and $d_i$ are denoted $d_i$ and $h_i$ respectively by Mather and

Jinks (1971, 1977). Experiments on crosses between pure lines and subsequent generations raised from them enable estimates to be made of the sums and sums of squares of the constants in eqn (4.25), as we shall see in the next chapter.

## Problems

1. In the general single locus, two allele case the gene frequencies of $B$ and $b$ are $p$ and $q(= 1 - p)$, and the genotypic values of $BB$, $Bb$, and $bb$ are $a$, $d$ and $-a$. Evaluate $A$ and $D$ as a function of the genotype under random mating; hence verify eqn (4.17).

2. Evaluate $A$ and $D$ as a function of the genotype in the above situation under complete inbreeding.

3. Evaluate the constants $a_1$, $a_2$, $d_1$, $d_2$, $aa_{12}$, $ad_{12}$, $ad_{21}$, and $dd_{12}$ defined in eqn (4.25) for the model of complementary gene action in Table 4.2 under complete inbreeding with equal gene frequencies.

4. Consider the following model of duplicate gene action:

|     | $BB$ | $Bb$ | $bb$ |
|-----|------|------|------|
| $CC$ | 1 | 1 | 1 |
| $Cc$ | 1 | 1 | 1 |
| $cc$ | 1 | 1 | 0 |

Find the genetic components of variance under random mating with equal gene frequencies at both loci.

5. Repeat Problem 3 for the model in Problem 4.

# 5 Biometric analysis of crosses between two inbred lines

*In a simple genetic experiment two pure lines, $P_1$ and $P_2$, are crossed to produce a hybrid generation, $F_1$; subsequent generations are obtained by selfing or sib-mating and by backcrossing to the parental or $F_1$ generations. To analyse this type of experiment we shall use the inbred decomposition of the genotypic value defined at the end of the last chapter. We suppose that the two parental lines differ at n autosomal loci affecting the trait under investigation, $P_1$ and $P_2$ being homozygous for the alleles $B_i$ and $b_i$ respectively at the ith locus. In the reference population each of the $2^n$ homozygotes which can be obtained by recombination and subsequent inbreeding occurs with equal frequency. We denote by $z_i$ the number of $B_i$ alleles in a particular individual ($z_i = 0$, 1, or 2); the vector $z = (z_1, z_2, \ldots, z_n)$ specifies the unordered genotype. We define the constants $a_i$, $d_i$, $aa_{ij}$, and so on as in eqn (4.25). The decomposition of the genotypic value is*

$$G(z) = m + \sum_i (z_i - 1)a_i + \sum_i z_i(2 - z_i)d_i + \sum_{i<j} (z_i - 1)(z_j - 1)aa_{ij}$$
$$+ \sum_{i \neq j} (z_i - 1)z_j(2 - z_j)ad_{ij} + \sum_{i<j} z_i(2 - z_i)z_j(2 - z_j)dd_{ij} + \cdots \tag{5.1}$$

*The theoretical mean and variance in any generation can be found by taking Expected values in (5.1), and the sums and sums of squares of the $a_i$s, $d_i$s, and so on can be estimated by comparing the observed and predicted means and variances in different generations. We shall begin by considering the statistical analysis of the mean values. For simplicity we shall ignore the possibility of genotype–environment interactions, though the model can easily be extended to include them.*

## Analysis of means

The decomposition of the genotypic value of an individual with genotype represented by the vector $z$ is given in eqn (5.1). In a specified generation $z_i$ is a random variable taking the values 0, 1, and 2 with probabilities $p_0, p_1$, and $p_2$ as discussed on page 30. Thus $(z_i - 1)$ takes values $-1$, 0, and 1 with these probabilities, while $z_i(2 - z_i)$ takes the values 0 and 1 with probabilities $(1 - p_1)$ and $p_1$. Hence

$$E(z_i - 1) = (p_2 - p_0)$$
$$E[z_i(2 - z_i)] = p_1. \tag{5.2}$$

Furthermore, in the absence of linkage values of $z_i$ at different loci are statistically independent. Taking the Expectation of (5.1) we find that

$$E(G) = m + (p_2 - p_0)a + p_1 d + (p_2 - p_0)^2 aa + (p_2 - p_0)p_1 ad + p_1^2 dd + \cdots$$
(5.3)

where

$$a = \sum_i a_i$$

$$d = \sum_i d_i$$
(5.4)

$$aa = \sum_{i<j} aa_{ij}$$

$$ad = \sum_{i \neq j} ad_{ij},$$

and so on. Thus information about the mean value in different generations can be used to estimate the constants $a$, $d$, $aa$, and so on defined in eqn (5.4).

The above result depends on the assumption that there is no linkage so that values of $z_i$ from different loci are statistically independent. If there is no epistasis the mean genotypic value is

$$E(G) = m + (p_2 - p_0) a + p_1 d$$
(5.5)

whether or not there is linkage, since linkage does not affect the Expected value of $z_i$ or $z_i^2$; but if there are epistatic interactions, linkage will contribute some extra terms to eqn (5.3) due to correlations between $z_i$s from different loci. The contribution from digenic interactions is

$$\sum_{i<j} \text{Cov}(z_i, z_j)aa_{ij} + \sum_{i \neq j} \text{Cov}(z_i, z_j(2 - z_j)) ad_{ij}$$

$$+ \sum_{i<j} \text{Cov}(z_i(2 - z_i), z_j(2 - z_j)) dd_{ij}$$
(5.6)

with similar contributions from interactions of higher orders. The covariances in eqn (5.6) can be evaluated from the results in Tables 3.2 and 3.3. For example, we find in the $F_k$ generation that

$$\text{Cov}(z_i, z_j) = (\tfrac{1}{2} - r_{ij})[1 - (\tfrac{1}{2} - r_{ij})^{k-1}]/(\tfrac{1}{2} + r_{ij})$$
$$\text{Cov}(z_i, z_j(2 - z_j)) = 0$$
(5.7)
$$\text{Cov}(z_i(2 - z_i), z_j(2 - z_j)) = [\tfrac{1}{2} - r_{ij}(1 - r_{ij})]^{k-1} - (\tfrac{1}{2})^{2(k-1)}.$$

We shall now illustrate how the model (5.3) can be fitted to data. Table 5.1, taken from Mather and Jinks (1977), shows the mean height of two parental varieties of tobacco and the $F_1$, $F_2$, and backcross generations derived from crossing them. The sample size was deliberately varied with the generation in order to give more information about the more variable segregating generations. All plants were individually randomized, and the different

TABLE 5.1

*Mean height (cm) in a cross between two varieties of tobacco (Mather and Jinks 1977)*

| Generation | No. of plants | Mean height ($\bar{y}$) | Var($\bar{y}$) | \multicolumn{6}{c}{Expected value} | | | | | |
|---|---|---|---|---|---|---|---|---|---|
| | | | | $m$ | $a$ | $d$ | $aa$ | $ad$ | $dd$ |
| $P_1$ | 20 | 116.30 | 1.00 | 1 | 1 | 0 | 1 | 0 | 0 |
| $P_2$ | 20 | 98.45 | 1.45 | 1 | $-1$ | 0 | 1 | 0 | 0 |
| $F_1$ | 60 | 117.68 | 0.97 | 1 | 0 | 1 | 0 | 0 | 1 |
| $F_2$ | 160 | 111.78 | 0.49 | 1 | 0 | $\frac{1}{2}$ | 0 | 0 | $\frac{1}{4}$ |
| $B_1$ | 120 | 116.00 | 0.49 | 1 | $\frac{1}{2}$ | $\frac{1}{2}$ | $\frac{1}{4}$ | $\frac{1}{4}$ | $\frac{1}{4}$ |
| $B_2$ | 120 | 109.16 | 0.61 | 1 | $-\frac{1}{2}$ | $\frac{1}{2}$ | $\frac{1}{4}$ | $-\frac{1}{4}$ | $\frac{1}{4}$ |

generations were grown in the same year, so that no complications from geno-type–environment interaction arise. The sampling variances, $\text{Var}(\bar{y})$, have been estimated by dividing the estimated variance within each generation by the sample size; these estimated sampling variances will be treated as if they are known without error. It has thus been assumed that different plants in the same generation are statistically independent; it will also be assumed that the means in the different generations are independent. These assumptions are valid for the six generations given, but they would break down in subsequent generations unless all the plants were unrelated to each other. The table also shows the predicted model (5.3) taking into account digenic interactions.

The six parameters in the model can be estimated from the six observed means by equating observed with Expected values. The standard errors of the estimates are easily found, since the latter are linear functions of the observed means. The estimates with their standard errors are shown below:

$$\hat{m} = 104.18 \pm 3.58 \quad \hat{aa} = \;\;\;3.20 \pm 3.50$$
$$\hat{a} = \;\;\;8.92 \pm 0.78 \quad \hat{ad} = -4.19 \pm 2.62$$
$$\hat{d} = \;\;16.92 \pm 8.80 \quad \hat{dd} = -3.41 \pm 5.64$$

None of the interactions is significantly different from zero. It therefore seems sensible to fit a model without interactions, taking into account only $m$, $a$, and $d$. To do this we use the method of weighted least squares since the observed means have different sampling variances. Suppose in general that the vector of observations, $\mathbf{y}$, can be represented by the linear model

$$\mathbf{y} = \mathbf{X}\boldsymbol{\beta} + \mathbf{e} \tag{5.8}$$

where $\mathbf{X}$ is a matrix of known coefficients, $\boldsymbol{\beta}$ is a vector of parameters to be estimated, and $\mathbf{e}$ is a vector of random errors with known variance–covariance matrix $\mathbf{V} = \text{Var}(\mathbf{e})$. (In the general case $\mathbf{V}$ may only be known up to a multiplicative constant, which must be estimated from the data, but we need not consider this complication here.) The method of weighted least squares is to find the estimates $\hat{\beta}$ which minimize the weighted sum of squares

$$(\mathbf{y} - \mathbf{X}\beta)^T \mathbf{V}^{-1} (\mathbf{y} - \mathbf{X}\beta). \tag{5.9}$$

The estimates are

$$\hat{\beta} = (\mathbf{X}^T \mathbf{V}^{-1} \mathbf{X})^{-1} \mathbf{X}^T \mathbf{V}^{-1} \mathbf{y}. \tag{5.10}$$

The variance–covariance matrix of these estimates is

$$\mathrm{Var}(\hat{\beta}) = (\mathbf{X}^T \mathbf{V}^{-1} \mathbf{X})^{-1}. \tag{5.11}$$

If we substitute the estimates (5.10) into (5.9), the resulting sum of squares has a chi-square distribution under normality with $n - p$ degrees of freedom, where $n$ observations have been made and $p$ parameters estimated. This can be used to test the goodness of fit of the model. This sum of squares is

$$\mathbf{y}^T \mathbf{V}^{-1} \mathbf{y} - \hat{\beta}^T \mathbf{X}^T \mathbf{V}^{-1} \mathbf{y}. \tag{5.12}$$

(Weighted least squares is equivalent to ordinary least squares after premultiplying both sides of (5.8) by $\mathbf{V}^{-1/2}$ to give the model

$$(\mathbf{V}^{-1/2} \mathbf{y}) = (\mathbf{V}^{-1/2} \mathbf{X})\beta + \mathbf{V}^{-1/2} \mathbf{e}.$$

Note that $\mathrm{Var}(\mathbf{V}^{-1/2} \mathbf{e}) = \mathbf{I}$, so that the errors in the transformed model are uncorrelated with unit variance. For further details see Draper and Smith (1966).)

In the present context $\mathbf{y}$ is the vector of means in the third column of Table 5.1; $\mathbf{X}$ is the $6 \times 3$ matrix of the coefficients of the parameters $m$, $a$, and $d$; $\beta$ is the vector of these parameters; and $\mathbf{V}$ is the diagonal matrix of the variances $\mathrm{Var}(\bar{y})$. The estimates of the parameters with their standard errors are:

$$\hat{m} = 107.32 \pm 0.67$$
$$\hat{a} = \phantom{0}8.21 \pm 0.63$$
$$\hat{d} = \phantom{0}10.06 \pm 1.23$$

The residual sum of squares (5.12) is 3.8, which is not significant as a chi-square value with three degrees of freedom. Thus there is no reason to suspect epistatic interactions in this cross. It will be seen that much more accurate estimates of the dominance and additive components can be obtained if it is assumed that there is no epistasis. The simplest interpretation of this experiment is that $P_1$ contains several genes for tallness which are completely dominant to their alleles in $P_2$. It must be remembered however that $a$ and $d$ are the sums of the *signed* quantities $a_i$ and $d_i$ over all loci. Since the $a_i$s and $d_i$s may vary in both sign and magnitude between loci, specified values of $a$ and $d$ are consistent with many different genetic models.

Epistatic interactions can sometimes be removed by a suitable transformation. Table 5.2 shows the result of a cross between two varieties of tomato which differed markedly in the size of the individual fruits. When this character is measured on an arithmetic scale there is marked epistasis, but this is removed by a logarithmic transformation. This transformation also serves to stabilize the variance in the non-segregating generations. Interpretation of these data is greatly simplified by a logarithmic transformation.

TABLE 5.2

*Mean weight per loculus of fruit and its standard error in a cross between two varieties of tomato (Powers 1951)*

| Generation | Arithmetic scale | Logarithmic scale |
|---|---|---|
| $P_1$ | $10.36 \pm 0.581$ | $0.977 \pm 0.027$ |
| $P_2$ | $0.45 \pm 0.017$ | $-0.364 \pm 0.018$ |
| $F_1$ | $2.33 \pm 0.130$ | $0.335 \pm 0.027$ |
| $F_2$ | $2.12 \pm 0.105$ | $0.273 \pm 0.015$ |
| $B_1$ | $4.82 \pm 0.253$ | $0.636 \pm 0.017$ |
| $B_2$ | $0.97 \pm 0.045$ | $-0.051 \pm 0.015$ |
| Chi-square with 3 d.f. testing for epistasis | 96.6 | 5.7 |

## Analysis of variances

Suppose that there is no epistasis so that the genotypic value is

$$G(\mathbf{z}) = m + \sum_i (z_i - 1)a_i + \sum_i z_i(2 - z_i)d_i. \tag{5.13}$$

The variables $(z_i - 1)$ and $z_i(2 - z_i)$ take the values $(-1, 0)$, $(0, 1)$, and $(1, 0)$ with probabilities $p_0, p_1$, and $p_2$ respectively. We compute

$$\mathrm{Var}(z_i - 1) = (1 - p_1) - (p_2 - p_0)^2$$
$$\mathrm{Var}(z_i(2 - z_i)) = p_1(1 - p_1) \tag{5.14}$$
$$\mathrm{Cov}((z_i - 1), z_i(2 - z_i)) = -p_1(p_2 - p_0).$$

We shall assume for the time being that there is no linkage so that values of these variables at different loci are independent and hence uncorrelated. We define

$$\alpha = \sum a_i^2$$
$$\delta = \sum d_i^2 \tag{5.15}$$
$$\gamma = \sum a_i d_i.$$

We now find from (5.13) that the variance of $G$ is

$$\mathrm{Var}(G) = [(1 - p_1) - (p_2 - p_0)^2]\alpha + p_1(1 - p_1)\delta - 2p_1(p_2 - p_0)\gamma. \tag{5.16}$$

The final term, which involves cross-products of additive and dominance components and which is not very easy to interpret, only occurs in the back-cross generations and can be eliminated by averaging the variances of complementary generations (such as $B_1$ and $B_2$).

TABLE 5.3

*Variance of height in a cross between two varieties of tobacco*

| Generation | No. of plants | Variance | Expected genetic variance | | |
|---|---|---|---|---|---|
| | | | $\alpha$ | $\delta$ | $\gamma$ |
| $P_1$ | 20 | 20.7 | — | — | — |
| $P_2$ | 20 | 29.0 | — | — | — |
| $F_1$ | 60 | 57.4 | — | — | — |
| $F_2$ | 160 | 77.7 | $\frac{1}{2}$ | $\frac{1}{4}$ | — |
| $B_1$ | 120 | 59.5 | $\frac{1}{4}$ | $\frac{1}{4}$ | $-\frac{1}{2}$ |
| $B_2$ | 120 | 66.2 | $\frac{1}{4}$ | $\frac{1}{4}$ | $\frac{1}{2}$ |

To illustrate the estimation of the parameters defined in (5.15), Table 5.3 shows the variance of height for the cross whose mean values are given in Table 5.1. The variance in $F_1$ is significantly larger than the parental variances, which suggests that heterozygosity increases the environmental variance; we shall assume that in any generation the observed variance contains an environmental contribution which can be estimated as $p_0\ \mathrm{Var}(P_2) + p_1\ \mathrm{Var}(F_1) + p_2\ \mathrm{Var}(P_0)$. The genetic parameters can be estimated by equating observed and Expected values of the segregating generations. The estimates with their standard errors are

$$\hat{a} = 4\ \mathrm{Var}(F_2) - 2\ \mathrm{Var}\ (B_1) - 2\ \mathrm{Var}(B_2) = 59.4 \pm 41.8$$

$$\hat{\delta} = 4\ \mathrm{Var}(B_1) + 4\ \mathrm{Var}(B_2) - 4\ \mathrm{Var}(F_2) - \mathrm{Var}(P_1) - \mathrm{Var}(P_2)$$
$$- 2\ \mathrm{Var}(P_1) = 27.5 \pm 62.7$$

$$\hat{\gamma} = \mathrm{Var}(B_2) - \mathrm{Var}(B_1) - \tfrac{1}{2}\mathrm{Var}(P_2) + \tfrac{1}{2}\mathrm{Var}(P_1) = 2.6 \pm 12.9.$$

The standard errors have been estimated by using the fact that, if $M$ is a mean square based on $f$ degrees of freedom then, under normality, $\mathrm{Var}\ (m) = 2E^2(M)/f$, which can be estimated by $2M^2/f$. (A refinement is to use $2M^2/(f + 2)$; this is an unbiased estimate since $E(M^2) = (f + 2)E^2(M)/f$.) For example,

$$\mathrm{Var}(\hat{a}) = 4^2 \times 2 \times \frac{77.7^2}{159} + 2^2 \times 2 \times \frac{59.5^2}{119} + 2^2 \times 2 \times \frac{66.2^2}{119} = 1748.$$

The standard errors are too large to allow reliable conclusions to be drawn.

Even if the above experiment were repeated on a larger scale it would still not be possible to test the validity of the model since all the degrees of freedom have been used in estimating the parameters. It is therefore desirable to use further generations, such as $F_3$ and $F_4$ or backcrosses involving $F_2$. The predicted genetic variance in any generation in the absence of epistasis and linkage is given by eqn (5.16) but with hierarchical mating systems it is often more meaningful to partition this variance into components representing the variances within and between families. Consider the $F_4$ generation as an example. Write $z_4$ for the genotype of an individual in $F_4$, and $z_3$ and $z_2$ for the genotypes of his ancestors in $F_3$ and $F_2$. We can write the deviation of the genotypic value, $G(z_4)$, from its mean value as

$$G - E(G) = [G - E(G|z_3)] + [E(G|z_3) - E(G|z_2)] \\ + [E(G|z_2) - E(G)] \tag{5.17}$$

where $E(G|z_3)$ is the mean value among individuals with the same parent, and $E(G|z_2)$ is the mean value among individuals with the same grandparent. Thus the first term on the right-hand side represents the deviation within families, the second term is the deviation between family means within superfamilies, and the third term is the deviation between means of superfamilies. Squaring both sides and taking Expected values over $z_4$, $z_3$, and $z_2$ in turn, we find that the cross-product terms vanish so that

$$\text{Var}(G) = \underset{z_3}{E} \text{Var}(G|z_3) + \underset{z_2}{E} \text{Var}[E(G|z_3)|z_2] + \text{Var } E(G|z_2). \tag{5.18}$$

This argument goes through unchanged in the presence of linkage and epistasis. Adopting the standard notation for these quantities, we write

$$V_{F_4} = V_{3F_4} + V_{2F_4} + V_{1F_4}. \tag{5.19}$$

$V_{3F_4}$ is the average variance within families, $V_{2F_4}$ is the average variance between family means within superfamilies, and $V_{1F_4}$ is the variance between superfamily means.

The decomposition can be extended to any generation under continued selfing. In $F_k$ we write

$$V_{F_k} = V_{k-1F_k} + V_{k-2F_k} + \cdots + V_{1F_k} \\ V_{jF_k} = \underset{z_j}{E} \text{Var}[E(G|z_{j+1})|z_j]. \tag{5.20}$$

Let $u_i = z_i - 1$ and $v_i = z_i(2 - z_i)$, and write $u_i(j)$ and $v_i(j)$ for the values of these variables in $F_j$. then

$$E(G|z_{j+1}) = m + \sum_i u_i(j+1)a_i + (\tfrac{1}{2})^{k-j-1} \sum_i v_i(j+1)d_i \tag{5.21}$$

since

$$E(u_i(k)|z_i(j)) = u_i(j) \\ E(v_i(k)|z_i(j)) = (\tfrac{1}{2})^{k-j} v_i(j). \tag{5.22}$$

We now compute the variance of eqn (5.21) conditional on $z(j)$. We observe that

$$\text{Var}(u_i(j+1)|z_i(j)) = \tfrac{1}{2} \text{ when } z_i(j) = 1 \\ \qquad\qquad\qquad\qquad 0 \text{ otherwise}$$

$$\text{Var}(v_i(j+1)|z_i(j)) = \tfrac{1}{4} \text{ when } z_i(j) = 1 \\ \qquad\qquad\qquad\qquad 0 \text{ otherwise} \tag{5.23}$$

$$\text{Cov}(u_i(j+1), v_i(j+1)|z_i(j)) = 0.$$

All covariances between variables from different loci are zero in the absence of linkage. Since the probability that $z_i(j) = 1$ is $(\frac{1}{2})^{j-1}$ independently at each locus, we eventually find that

$$V_{jF_k} = (\tfrac{1}{2})^j \alpha + (\tfrac{1}{2})^{2k-j-1}\delta. \tag{5.24}$$

In particular, the within family variance is

$$V_{(k-1)F_k} = (\tfrac{1}{2})^{k-1} \alpha + (\tfrac{1}{2})^k \delta. \tag{5.25}$$

The total variance is

$$V_{F_k} = (1 - (\tfrac{1}{2})^{k-1})\alpha + (\tfrac{1}{2})^{k-1} (1 - (\tfrac{1}{2})^{k-1})\delta. \tag{5.26}$$

in agreement with eqn (5.16).

It may also be of interest to consider the covariances between successive generations. Consider in particular the covariance between the values of an individual in $F_{k-1}$ and his offspring in $F_k$. We first compute the Expected value of the offspring conditional on their parent from eqn (5.21):

$$E(G(\mathbf{z}_k)|\mathbf{z}_{k-1}) = m + \sum u_i(k-1)a_i + \tfrac{1}{2}\sum v_i(k-1)d_i. \tag{5.27}$$

$G(\mathbf{z}_{k-1})$ can be expressed in the same form as (5.27), but without the coefficient $\frac{1}{2}$. By using eqn (5.14) we find that the covariance of these two quantities, averaged over the distribution of $\mathbf{z}_{k-1}$, is

$$(1 - (\tfrac{1}{2})^{k-2})(\alpha + (\tfrac{1}{2})^{k-1} \delta). \tag{5.28}$$

Similar arguments can be used to find the decomposition of the variance under repeated backcrossing and other mating systems, but we will not pursue the details here.

To illustrate the application of these results we shall describe an experiment discussed by Hayman (1960) on height (in inches) in a cross of two varieties of tobacco. Data were obtained on the two inbred parents ($P_1$ and $P_2$) and on their $F_1$, $F_2$, $F_3$, and $F_4$ progeny. All the generations were grown together in two randomized blocks in 1955. Each block contained 150 plots, and five plants were grown in each plot. The plots in each block were allocated as follows:

5 plots of $P_1$, each plot with 5 plants
5 plots of $P_2$, each plot with 5 plants
10 plots of $F_1$, each plot with 5 plants, obtained by crossing $P_1$ and $P_2$ in 1954
10 plots of $F_2$, each plot with 5 plants, obtained by crossing $P_1$ and $P_2$ in 1953 and selfing the $F_1$ in 1954
20 plots of $F_3$, each plot with 5 plants representing a single $F_3$ family, obtained by crossing $P_1$ and $P_2$ in 1952, selfing 20 $F_2$ plants in 1954, and raising 5 seedlings from each plant
100 plots of $F_4$, each plot with 5 plants representing a single $F_4$ family, obtained by crossing $P_1$ and $P_2$ in 1951, selfing 20 $F_2$ plants in 1953, raising 5 seedlings from each in 1954, selfing all 100 plants and raising 5 seedlings from each in 1955; the 100 plots represent 20 superfamilies.

The advantages of this design are (i) that all plants are grown under similar conditions in the same year so that the results are not complicated by genotype–year interaction, and (ii) that plants in different generations are genetically unrelated to each other. In $F_1$ and subsequent generations equal numbers of plots were allocated to reciprocal crosses (depending on whether $P_1$ or $P_2$ was the pollen parent) and these reciprocal crosses were kept separate in the analysis.

TABLE 5.4

*Analysis of variance of height in a cross between two varieties of tobacco* (*Hayman 1960*)

| Variance component | Observed value | Degrees of freedom | Expected value | | | |
|---|---|---|---|---|---|---|
| | | | $\alpha$ | $\delta$ | $V_{Eb}$ | $V_{Ew}$ |
| $M_{1F_2}$ | 69.29 | 80 | 0.5 | 0.25 | — | 1 |
| $M_{1F_3}$ | 43.12 | 36 | 0.55 | 0.0875 | 1 | 0.2 |
| $M_{2F_3}$ | 36.66 | 160 | 0.25 | 0.125 | — | 1 |
| $M_{1F_4}$ | 67.84 | 36 | 0.555 | 0.024375 | 0.2 | 0.04 |
| $M_{2F_4}$ | 41.29 | 153† | 0.275 | 0.04375 | 1 | 0.2 |
| $M_{3F_4}$ | 26.47 | 770† | 0.125 | 0.0625 | — | 1 |
| $M_{Eb}$ | 14.06 | 32 | — | — | 1 | 0.2 |
| $M_{Ew}$ | 12.95 | 160 | — | — | — | 1 |
| Estimate ± standard error | | | $97.5 \pm 19.3$ | $12.6 \pm 48.8$ | $10.8 \pm 3.1$ | $13.2 \pm 1.4$ |

$\chi^2 = 3.7$ with four degrees of freedom.

† Degrees of freedom reduced due to loss of a few plants or plots.

The analysis of the data is summarized in Table 5.4. The observed variance components are denoted by $M$ (with a subscript), to distinguish them from the corresponding theoretical variances, denoted by $V$. There are two sources of environmental variability, between plants within plots and between plots within blocks. We denote the corresponding theoretical variances by $V_{Ew}$ and $V_{Eb}$ respectively; they can be estimated from the observed variance components $M_{Ew}$ and $M_{Eb}$ at the bottom of the table. $M_{Ew}$ is the observed variance between plants within $P_1$, $P_2$, and $F_1$ plots and it has 160 degrees of freedom since there are 40 such plots each with four degrees of freedom. There was no evidence of heterogeneity of variance between $P_1$, $P_2$, and $F_1$ plots. $M_{Eb}$ is the observed variance between the means of replicate $P_1$, $P_2$, and $F_1$ plots within each block. It has 32 degrees of freedom since $P_1$, $P_2$, and each of the reciprocal $F_1$s contributes four degrees of freedom in each block. The Expected values of these components are shown in the table. Note that $M_{Ew}$ is calculated per plant, while $M_{Eb}$ is calculated per plot.

We turn now to the genetic variance components. $M_{1F_2}$ is the within-plot variance between $F_2$ plants. It has 80 degrees of freedom since there are 20 $F_2$ plots each with four degrees of freedom. Its Expected value is

$$E(M_{1F_2}) = V_{1F_2} + V_{Ew}.$$

(The information about $V_{1F_2}$ contained in the plot means has been ignored.) $M_{1F_3}$ is the variance among the means of $F_3$ plots; it has $36 = 9 \times 4$ degrees of freedom since there are ten $F_3$ plots in each block from each reciprocal. Its Expected value is

$$E(M_{1F_3}) = V_{1F_3} + V_{Eb} + \tfrac{1}{5}(V_{2F_3} + V_{Ew})$$

since the random effects model of the analysis of variance is clearly applicable here. $M_{2F_3}$ is the variance within $F_3$ plots and has Expected value $V_{2F_3} + V_{Ew}$. $M_{1F_4}$ is the variance between the means of $F_4$ superfamilies, each derived from the same $F_2$ ancestor; $M_{2F_4}$ is the variance between the means of $F_4$ families, derived from the same $F_3$ ancestor, within superfamilies; $M_{3F_4}$ is the variance within $F_4$ families, that is to say within $F_4$ plots. Their Expected values are

$$E(M_{1F_4}) = V_{1F_4} + \tfrac{1}{5}(V_{2F_4} + V_{Eb}) + \tfrac{1}{25}(V_{3F_4} + V_{Ew})$$

$$E(M_{2F_4}) = V_{2F_4} + V_{Eb} + \tfrac{1}{5}(V_{3F_4} + V_{Ew})$$

$$E(M_{3F_4}) = V_{3F_4} + V_{Ew}.$$

Finally the genetic variances $V_{jF_k}$ can be written in terms of $\alpha$ and $\delta$ from eqn (5.24), whence the Expected values in Table 5.4 are obtained.

Thus the Expected values of the eight variance components in Table 5.4 have been expressed as linear functions of the four parameters $\alpha$, $\delta$, $V_{Eb}$, and $V_{Ew}$. From the design of the experiment these variance components are independently distributed, and their sampling variances can be estimated as $2M^2/f$, assuming normality. The parameters can now be estimated by weighted least squares, as described in the previous section. A refinement is to re-estimate the sampling variances as $2\hat{M}^2/f$, where $\hat{M}$ is the Expected value of $M$, using the estimated parameter values just obtained; this procedure can be repeated iteratively until the estimates converge to a steady value. The rationale of this method is that, if the model fits, $\hat{M}$ will be a more reliable estimate of $E(M)$ than $M$ itself. The estimates in Table 5.4 were obtained in this way. There is a satisfactory goodness of fit from the chi-square value, but the standard error of $\hat{\delta}$ is discouragingly large.

*Linkage*

The above argument assumes that there is no linkage so that different loci are independently distributed. In the presence of linkage (but still without epistasis) the following extra terms must be added to the expression for the total variance in eqn (5.16):

$$2 \sum_{h<i} \mathrm{Cov}(u_h, u_i) a_h a_i + 2 \sum_{h \neq i} \mathrm{Cov}(u_h, v_i) a_h d_i$$

$$+ 2 \sum_{h<i} \mathrm{Cov}(v_h, v_i) d_h d_i. \tag{5.29}$$

The covariances in eqn (5.29) can be evaluated from the results in Tables 3.2 and 3.3. They are given for $F_k$ in eqn (5.7).

To extend the argument used in deriving the components of variance (5.24) to include linkage, we first derive the covariances between $us$ and $vs$ from two different loci in $F_{j+1}$ given the genotype in $F_j$. These covariances are zero unless both loci are segregating in $F_j$, which means a double heterozygote. In this case we must distinguish whether the double heterozygote is in coupling or repulsion. By an obvious calculation we find that

$$\mathrm{Cov}(u_h(j+1), u_i(j+1)|z_h(j) = z_i(j) = 1) = \pm(\tfrac{1}{2} - r_{hi})$$

$$\mathrm{Cov}(u_h(j+1), v_i(j+1)|z_h(j) = z_i(j) = 1) = 0 \qquad (5.30)$$

$$\mathrm{Cov}(v_h(j+1), v_i(j+1)|z_h(j) = z_i(j) = 1) = (\tfrac{1}{2} - r_{hi})^2.$$

The sign of the first term depends on whether the double heterozygote is in coupling or repulsion. Averaging over the frequencies of coupling and repulsion double heterozygotes in $F_j$ given in Table 3.3 we find that

$$V_{jF_k} = (\tfrac{1}{2})^j \alpha_j + (\tfrac{1}{2})^{2k-j-1} \delta_j, \qquad (5.31)$$

where

$$\alpha_j = \sum_i a_i^2 + 2 \sum_{h<i} (1 - 2r_{hi})^j a_h a_i$$

$$\delta_j = \sum_i d_i^2 + 2 \sum_{h<i} [1 - 2r_{hi}(1 - r_{hi})]^{j-1} (1 - 2r_{hi})^2 d_h d_i. \qquad (5.32)$$

Given experimental results like those in Table 5.4 the eight parameters $\alpha_j$, $\delta_j$ ($j = 1, 2, 3$), $V_{Eb}$, and $V_{Ew}$ can be estimated by equating observed and Expected values, though no degrees of freedom remain to test goodness of fit. A test of goodness of fit can be obtained if backcross generations are included in the experimental design. For further details see Mather and Jinks (1971).

### *Epistasis

The preceding analysis presupposes that there are no epistatic interactions between loci. The presence of epistasis can be tested from the analysis of the means as described in the previous section. If there is reason to suspect epistasis, it can be included in the analysis of the variances, but the results become so complicated that it seems doubtful whether it is worthwhile to pursue the analysis; in any case it would only be possible, because of the complexity of the results, to include digenic interactions, and there seems little reason why higher interactions should not be important if epistasis is present at all. To illustrate the theory, we shall derive the variance components $V_{jF_k}$ with digenic interactions in the absence of linkage.

The decomposition of the genotypic value is given in eqn (5.1). The extension of eqn (5.21) is

$$E(G|z_{j+1}) = m + \sum_i u_i(j+1)a_i + (\tfrac{1}{2})^{k-j-1} \sum_i v_i(j+1)d_i$$

$$+ \sum_{h<i} u_h(j+1)u_i(j+1)aa_{hi}$$

$$+ (\tfrac{1}{2})^{k-j-1} \sum_{h \neq i} u_h(j+1)v_i(j+1)ad_{hi} \tag{5.33}$$

$$+ (\tfrac{1}{2})^{2(k-j-1)} \sum_{h<i} v_h(j+1)v_i(j+1)dd_{hi}$$

since loci behave independently in the absence of linkage. The variances and covariances of the individual terms in eqn (5.33) conditional on the genotype in $F_j$ are easily found by direct calculation. Averaging the variance of eqn (5.33) over the genotypic distribution in $F_j$ we eventually find that

$$V_{JF_k} = (\tfrac{1}{2})^j \sum_i a_i^2 + (\tfrac{1}{2})^{2k-j-1} \sum_i d_i^2 + [(\tfrac{1}{2})^{j-1} - 3(\tfrac{1}{2})^{2j}] \sum_{h<i} aa_{hi}^2$$

$$+ (\tfrac{1}{2})^{2k-j-1} \sum_{h \neq i} ad_{hi}^2 + 3(\tfrac{1}{2})^{4k-2j-2} \sum_{h<i} dd_{hi}^2 + (\tfrac{1}{2})^{k+j-2} \sum_{h \neq i} a_h \, ad_{hi}$$

$$+ (\tfrac{1}{2})^{3k-j-3} \sum_{h \neq i} d_h \, dd_{hi} + (\tfrac{1}{2})^{2k+j-2} \sum_{h \neq i \neq l} ad_{hi} \, ad_{hl} \tag{5.34}$$

$$+ (\tfrac{1}{2})^{4k-j-4} \sum dd_{hi} \, dd_{lm}.$$

(The last sum is over $h < i$, $l < m$, together with exactly one of the following conditions: $h = l$ or $h = m$ or $i = l$ or $i = m$.)


## The North Carolina 3 design

The main problem encountered in this section has been the low precision of the estimates obtained, particularly that of the dominance component $\delta$. Comstock and Robinson (1952) described three experimental designs to obtain more accurate estimates of the degree of dominance; they have become known as the North Carolina designs because they were developed and extensively used in that state. The most efficient of these designs for estimating dominance is the third design, which will be discussed now.

In the North Carolina 3 design an $F_2$ generation is first raised following a cross between two inbred lines; $F_2$ plants are then crossed back to the original inbred lines. A number of $F_2$ plants are chosen, each of them is crossed to each of the two parental lines, and a number of plants is raised from each mating.

Suppose that $s$ $F_2$ plants are chosen, and that $r$ plants (or replicate plots) are measured from each mating type. Write $y_{ijk}$ for the $k$th observation on the $i$th $F_2$ plant mated to the $j$th parental line ($i = 1, \ldots, s; j = 1, 2; k = 1, \ldots, r$). This is a crossed design which, in the absence of subdivision into blocks, gives rise to the analysis of variance in Table 5.5. $\alpha$ and $\delta$ can be estimated by equating observed and Expected values; the variances and covariance of these estimates are simply calculated since the estimates are linear functions of the mean squares.

TABLE 5.5

*Analysis of variance of the North Carolina 3 design*

| Source of variation | Sum of squares | Degrees of freedom | Expected mean square |
|---|---|---|---|
| Between inbred lines | $\frac{1}{2}rs(y_{.1.} - y_{.2.})^2$ | 1 | |
| Between $F_2$ parents | $2r \sum (y_{i..} - y_{...})^2$ | $(s-1)$ | $\frac{1}{4}r\alpha + V_e$ |
| Lines × $F_2$ parents | $r \sum (y_{ij.} - y_{i..} - y_{.j.} + y_{...})^2$ | $(s-1)$ | $\frac{1}{4}r\delta + V_e$ |
| Residual error | $\sum (y_{ijk} - y_{ij.})^2$ | $2s(r-1)$ | $V_e$ |

NB $V_e$ includes part of the genetic variance as well as the environmental variance.

Comstock and Robinson (1952) were mainly interested in estimating the average level of dominance. The level of dominance at a particular locus is measured by $|d_i/a_i|$, with values $>1$, $=1$, or $<1$ representing overdominance, complete dominance, and underdominance respectively; thus $(\delta/\alpha)^{1/2}$ can be interpreted as a measure of the average level of dominance. As we shall see in Chapter 7 this quantity is important in interpreting the important phenomena of inbreeding depression and hybrid vigour.

Estimates of the average level of dominance may be biased by the presence of linkage. In deriving the Expected mean squares in Table 5.5 it has been assumed that there is linkage equilibrium in the $F_2$ generation. This will be true for unlinked loci, but will not usually be true for linked loci. If there is directional dominance, so that all the $d_i$s have the same sign, linkage will tend to inflate the estimated value of $\delta/\alpha$. To see the reason for this, consider a pair of loci with alleles $B$, $b$ and $C$, $c$ respectively; let the contributions of $BB$, $Bb$, and $bb$ be 1, $d$, and $-1$ respectively, and suppose that the $C$ locus behaves in the same way. If the loci are closely linked in coupling in the parental lines, so that $P_1$ is $BC/BC$ and $P_2$ is $bc/bc$, this pair of loci will for the first few generations, before much crossing over has occurred, behave like a single locus with values 2, $2d$, and $-2$ for the genotypes $BC/BC$, $BC/bc$, and $bc/bc$; this will lead to increased variability, but there is no change in the apparent level of dominance. However, if they are linked in repulsion, so that $P_1$ is $Bc/Bc$ while $P_2$ is $bC/bC$, they will behave like a single locus with values 0, $2d$, and 0 for the genotypes $Bc/Bc$, $Bc/bC$, and $bC/bC$; in this case there is a great increase in the apparent level of dominance. In general, $\alpha$ and $\delta$ in Table 5.5 should be replaced by $\alpha'$ and $\delta'$ defined as

$$\alpha' = \sum_i a_i^2 + \sum_{i \neq j} (1 - 2r_{ij})a_i\,a_j$$

$$\delta' = \sum_i d_i^2 + \sum_{i \neq j} (1 - 2r_{ij})d_i\,d_j,$$

(5.35)

where $r_{ij}$ is the recombination fraction between the $i$th and $j$th loci. If linkage occurs equally often in coupling and repulsion (as it would do by chance), then $a_i\,a_j$ is equally likely to be positive or negative, and the bias in $\alpha$ is likely to be small. On the other hand, with directional dominance $d_i d_j$ will always be positive and $\delta$ is likely to be overestimated.

Moll, Lindsey, and Robinson (1964) have investigated the importance of linkage in biasing estimates of the level of dominance in maize. They crossed two inbred lines and estimated $\alpha$ and $\delta$ from a North Carolina 3 design on the $F_2$. They then mated this $F_2$ generation at random for 6–10 generations to allow linkage disequilibrium to disappear, and re-estimated $\alpha$ and $\delta$ by backcrossing the randomly-mated plants to the inbred lines. They found that estimates of $\alpha$ changed little, but that estimates of $\delta$ and of the average level of dominance declined for many characters after several generations of random mating; the average level of dominance was sometimes greater than or about 1 when measured in $F_2$, but fell well below 1 after random mating. They concluded that linkage causes an upward bias in estimates of dominance obtained from $F_2$ populations, and that what appears to be overdominance is not true overdominance but is due to loci temporarily linked in repulsion; this phenomenon is called *associative overdominance*.

**Problems**

1. Estimate $m$, $a$, and $d$ from the log-transformed data of Table 5.2 on the assumption that there is no epistasis. Find the standard errors of these estimates. Verify the chi-square value for testing epistasis.

2. Mather and Jinks (1977) quote the following data on a cross of two varieties of tobacco:

| Generation | Mean height $\pm$ standard error |
|------------|----------------------------------|
| $P_1$ | $80.4 \pm 1.9$ |
| $P_2$ | $65.5 \pm 1.7$ |
| $F_1$ | $86.0 \pm 1.2$ |
| $F_2$ | $84.0 \pm 0.9$ |
| $B_1$ | $84.2 \pm 1.2$ |
| $B_2$ | $73.9 \pm 1.0$ |

Estimate the six parameters in a model including digenic interactions, together with their standard errors. Do you think that a model without interactions is likely to fit well?

3. If a character is determined by $n$ unlinked loci with equal additive effects, $a_i$, show that $n = a^2/\alpha$. This fact can be used to estimate $n$. What will be the effect on this estimate if the $a_i$s differ in magnitude or sign?

4. Estimate the number of loci for the data in Tables 5.1 and 5.3 by this method. Is this estimate consistent with the estimated value of $\gamma$?

5. In discussing the analysis of the data in Table 5.4, we remarked that the information about $V_{1F_2}$ contained in the plot means was ignored. How could this information be incorporated into the analysis?

6. Verify eqn (5.24). Verify the terms $E(M_{jF_k})$ in Table 5.4.

7. Verify the Expected mean squares in Table 5.5 and evaluate $V_e$.

8. Verify that the Expected mean squares in Table 5.5 remain valid in the presence of linkage if $\alpha$ and $\delta$ are replaced by $\alpha'$ and $\delta'$, defined in eqn (5.35).

# 6 Biometric analysis of outbred populations

*In this chapter we shall consider the analysis of genetic variability of a quantitative character in an outbred population in which random mating can be assumed. Such analysis is based on observing correlations between relatives. We shall begin by considering the simplest situation in which epistasis and other complications can be ignored.*

## The elementary model

We wish to derive the theoretical correlation between the genotypic values of two related individuals under the simplest possible assumptions. Consider first a character determined by a single autosomal locus; write the genotypes of the two individuals as $B'_p B'_m$ and $B''_p B''_m$ respectively, where the subscripts refer to genes of paternal and maternal origin. Decomposing the genotypic value under the random mating decomposition defined in Chapter 4 we can write

$$G' = m + (X'_1 + X'_2) + X'_{12} = m + A' + D'$$
$$G'' = m + (X''_1 + X''_2) + X''_{12} = m + A'' + D''.$$

We now consider covariances between the two individuals conditional on the number of pairs of identical genes. If there are no identical genes then the genotypic values, or any components of them, in the two individuals are independent and so uncorrelated. If there are two pairs of identical genes, then $G' = G''$, $A' = A''$, and $D' = D''$, so that

$$\mathrm{Cov}(G', G'') = V_G$$
$$\mathrm{Cov}(A', A'') = V_A$$
$$\mathrm{Cov}(D', D'') = V_D.$$

Note however that $A'$ and $D''$ are uncorrelated since it has already been shown that $A'$ and $D'$ are uncorrelated. Suppose finally that there is one pair of identical genes, and let this pair be the paternal gene in both individuals, so that $B'_p = B''_p$; the maternal genes will be statistically independent both of each other and of $B'_p$. In this case $X'_1 = X''_1$, but

$$E(X'_1 X'_2) = E(X'_1 X''_{12}) = E(X'_2 X''_{12}) = E(X'_{12} X''_{12}) = 0.$$

(To prove this result consider the conditional Expectations given $B'_p$, note that the variables are now pairwise independent, and that at least one of them has zero Expectation from eqn (4.12).) Since $\text{Var}(X_1') = \frac{1}{2}V_A$ it follows that

$$\text{Cov}(G', G'') = \text{Cov}(A', A'') = \frac{1}{2}V_A$$
$$\text{Cov}(A', D'') = \text{Cov}(D', D'') = 0.$$

The same result is obtained whichever pair of genes is assumed to be identical.

Averaging over the three cases of 0, 1, or 2 pairs of identical genes (which occur with probabilities $P_0$, $P_1$, and $P_2$), and writing $R = \frac{1}{2}P_1 + P_2$, we find that

$$\text{Cov}(G', G'') = RV_A + P_2V_D$$
$$\text{Cov}(A', A'') = RV_A$$
$$\text{Cov}(D', D'') = P_2V_D \tag{6.1}$$
$$\text{Cov}(A', D'') = 0.$$

$R$ is usually called the coefficient of relationship between the relatives; it is the correlation between their additive effects. $2R$ is the average number of pairs of identical genes per locus.

This result can be extended to the multi-locus case without epistasis. We write

$$G' = m + (A_1' + A_2' + \cdots + A_n') + (D_1' + D_2' + \cdots + D_n') = m + A' + D'$$
$$G'' = m + (A_1'' + A_2'' + \cdots + A_n'') + (D_1'' + D_2'' + \cdots + D_n'') = m + A'' + D''.$$

It can be shown by the same argument as before that

$$\text{Cov}(A_i', A_i'') = R \, \text{Var}(A_i)$$
$$\text{Cov}(D_i', D_i'') = P_2 \, \text{Var}(D_i)$$
$$\text{Cov}(A_i', D_i'') = 0.$$

Components from different loci, such as $A_i'$ and $A_j''$ for $i \neq j$, are independent since all genes which are not identical by descent are statistically independent under random mating. Eqn (6.1) follows by adding over loci.

The theoretical covariances for some common relatives are shown in Table 6.1. They were found by using (6.1) together with Table 3.5. If it is assumed that there are no environmental correlations then these are also the covariances between the phenotypic values. The phenotypic correlations are obtained by dividing by the phenotypic variance $V_Y$. These correlations depend on $V_A/V_Y$ and $V_D/V_Y$, which express the proportion of the phenotypic variance due to additive genetic variance and dominance respectively; the important ratio $V_A/V_Y$ is called the heritability, denoted by $h^2$. It is often more convenient to express the relationship between parent and child in terms of a regression rather than a correlation coefficient. Writing $Y_F$, $Y_M$, and $Y_C$ for the phenotypic values

TABLE 6.1

*Genetic covariances between relatives without epistasis*

| Relationship | Coefficient of | |
| --- | --- | --- |
| | $V_A$ | $V_D$ |
| Identical twins | 1 | 1 |
| Parent–child | 0.5 | 0 |
| Sibs | 0.5 | 0.25 |
| Grandparent–grandchild, half-sibs, uncle–nephew | 0.25 | 0 |
| Greatgrandparent–greatgrandchild, cousins | 0.125 | 0 |

of father, mother, and child, and assuming that the regression is linear, we find from the covariances in Table 6.1 that the regression of child on father is

$$E(Y_C|Y_F) = m + \tfrac{1}{2}h^2 (Y_F - m). \qquad (6.2)$$

The joint regression on both parents is

$$E(Y_C|Y_F, Y_M) = m + \tfrac{1}{2}h^2 (Y_F - m) + \tfrac{1}{2}h^2 (Y_M - m). \qquad (6.3)$$

Thus the regression coefficient on the mid-parental value, $\tfrac{1}{2}(Y_F + Y_M)$, is $h^2$. These regressions are discussed in more detail in the next chapter.

To illustrate the application of this theory, we shall discuss the inheritance of the fingerprint ridge count in man, which provides a good example of what happens in the absence of any complications. A series of family studies of this character are summarized by Holt (1968). The ridges on each finger are counted according to defined rules depending on the fingerprint pattern and are then summed for all ten fingers to give the total ridge count. The correlations between relatives are shown in Table 6.2, and the regression of child on mid-parent is shown graphically in Fig. 6.1. The parameters $h^2$ and $V_D/V_Y$ can be estimated from the observed values by weighted least squares, with one degree of freedom left to test the validity of the model, after combining the correlations for sibs and dizygotic twins; however, it is clear from inspection that the observed values are in good agreement with theory, with $h^2 = 0.95$, $V_D/V_Y = 0$.

TABLE 6.2

*Corelations between relatives for total fingerprint ridge count*

| Type of relationship | Observed value ± standard error | Theoretical value Coefficient of | |
| --- | --- | --- | --- |
| | | $h^2$ | $V_D/V_Y$ |
| Correlation between monozygotic twins | 0.95 ± 0.01 | 1 | 1 |
| Correlation between dizygotic twins | 0.49 ± 0.08 | $\tfrac{1}{2}$ | $\tfrac{1}{4}$ |
| Correlation between sibs | 0.50 ± 0.04 | $\tfrac{1}{2}$ | $\tfrac{1}{4}$ |
| Regression of child on mid-parent | 0.93 ± 0.04 | 1 | 0 |

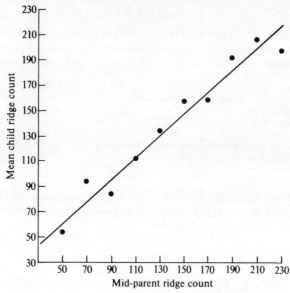

Fig. 6.1 Regression of child on mid-parent for fingerprint ridge count (Holt 1968).

It can be concluded that 95 per cent of the variability is of genetic and 5 per cent of environmental origin, and that there is completely additive gene action with no evidence of either dominance or epistasis.

*Environmental correlation*

The assumption has been made so far that the phenotypic correlation between relatives arises entirely from their genetic relationship, but it should always be kept in mind that they may also resemble each other because they share similar environmental circumstances. We may write the phenotypic values of the relatives as

$$Y' = G' + E'$$

$$Y'' = G'' + E''.$$

The phenotypic covariance is

$$\mathrm{Cov}(Y', Y'') = \mathrm{Cov}(G', G'') + \mathrm{Cov}(E', E'').$$

(Remember that $G$ and $E$ are by definition uncorrelated.) The environmental covariance arises from the fact that members of the same family are more likely to share similar environments than members of different families. It is likely to be most pronounced in full sibs, particularly if they belong to the same litter or clutch. For example, Johansson and Korkman (1951), in a study of growth rate in pigs, report a correlation of 0.21 among litter mates compared with 0.13 among full sibs that are not litter mates. In experimental situations, environmental correlation after birth can usually be avoided by randomization, and in man use can be made of the natural experiment provided by adopted children; comparison can be made between twins or sibs reared together and those reared apart, and between the resemblance of adopted children with their natural and their foster parents. Nevertheless, environmental correlation presents a serious problem in the interpretation of correlations between relatives and its possibility must always be remembered.

As remarked above, environmental correlation is likely to be most marked in litter mates. For example, twin data in cattle have turned out to be of little value for this reason. Thus the regression of daughter on dam for milk yield is about 0.15, so that in the absence of dominance, epistasis or environmental correlation one would expect the correlations between dizygotic and monozygotic twins to be about 0.15 and 0.30 respectively. The observed values are 0.67 for dizygotic and 0.88 for monozygotic twins (Johansson and Rendel 1968). The high value for monozygotic twins might be accounted for by dominance or epistasis, but this cannot explain the dizygotic correlation. It therefore seems likely that both these correlations have been grossly inflated by environmental correlation due to the fact that twins share a common intra-uterine environment before birth and are reared in similar circumstances after birth.

In the above case it seems likely that the environmental correlation is due to simultaneity of development and birth in a temporally fluctuating environment; sibs from different litters would not experience this correlation. In other cases the correlation is due to the maternal environment, which is shared by sibs from the same or from different litters. In this situation it may be appropriate to regard the character expressed in the children as being determined by the genotype of the mother; for a discussion of the inheritance of human birth weight from this point of view see Bulmer (1970).

## Statistical methods

In this section we shall consider some of the statistical methods which have been found useful in analysing correlations and regressions between relatives. The aim is not to provide a complete catalogue of statistical methodology in this area, which would require a book to itself, but to illustrate some of the problems which arise and how they can be tackled, in the hope that the general principles can be extended to other problems.

### Regression of offspring on parent

The resemblance between parent and child is usually measured by the regression of the child's value on the parental value. We shall assume that this regression is linear and that measurements in the two sexes have the same mean and variance; the latter can always be achieved by a preliminary standardization. If both parents have been measured and if the inheritance of the character is unaffected by the sex of the individual, after standardization if necessary, we shall usually want to consider the regression of the child's value on the mid-parental value; in the absence of epistasis and of environmental correlations between parent and child this gives a direct estimate of heritability. If the character can only be measured in one sex, or if there is a possibility of differences between the sexes in inheritance, we shall want to consider the regression on one parent only, for example the regression of daughters on mothers; twice the regression coefficient gives an estimate of heritability.

To illustrate the statistical problems which may arise we shall consider the regression on mid-parental value. Write $x_i$ for the average of the phenotypic values of the $i$th pair of parents, and suppose that each pair has one child with phenotypic value $y_i$. The statistical model is

$$y_i = m + \beta x_i + e_i \qquad (6.4)$$

where

$$\mathrm{Var}(e_i) = (1 - \tfrac{1}{2}\beta^2)V \qquad (6.5)$$

and where $\beta = h^2$, in the absence of epistasis or environmental correlation. (For the calculation of $\mathrm{Var}(e_i)$ see eqn (8.4).) The regression is estimated in the usual way by

$$\hat{\beta} = \sum (x_i - \bar{x})(y_i - \bar{y})/\sum (x_i - \bar{x})^2 \qquad (6.6)$$

with sampling variance

$$\mathrm{Var}(\hat{\beta}) = \mathrm{Var}(e)/\sum (x_i - \bar{x})^2; \qquad (6.7)$$

the residual variance $\mathrm{Var}(e)$ can be estimated from eqn (6.5) or from the residual mean square.

It should be noticed that the sampling variance is inversely proportional to the sum of squares of the $x_i$s. In experimental situations we can increase this sum of squares and hence decrease the sampling variance by choosing as parents individuals with extreme values and by mating them assortatively. This procedure will not affect the statistical model (6.4) or the residual error variance (6.5), though it will of course affect the statistical properties of subsequent generations. Selection of parents would of course affect the estimate (6.6) if the true regression were non-linear. It will also affect the correlation between parent and offspring, since it affects the variance of the offspring distribution. This is one reason for preferring regression to correlation methods.

A statistical problem arises if there are several children in each family, particularly if there are different numbers of children in different families. It would be wrong to treat children in the same family as having independent values, since they are full sibs. To get round this problem we might consider the regression of the mean value of the children in a family on the mid-parental value; but in this case we should give more weight to large than to small families, since their mean values are more accurate. We must therefore investigate the accuracy of family means.

The statistical model is

$$y_{ij} = m + a_i + \beta x_i + e_{ij}. \tag{6.8}$$

In this equation $y_{ij}$ is the value of the $j$th child of the $i$th pair, $m$ is the overall mean, $a_i$ is the deviation of the Expected value of children in this family from $m + \beta x_i$, and $e_{ij}$ is the deviation of a particular child in this family from the family mean due to variability between sibs. The $a_i$s and the $e_{ij}$s are uncorrelated random variables whose variances we shall now find. If $\rho$ is the correlation between full sibs, then under random mating

$$\text{Cov}(y_{ij}, y_{ik}) = \rho V = \text{Var}(a_i) + \tfrac{1}{2}\beta^2 V.$$
$$\text{Var}(y_{ij}) = \text{Var}(a_i) + \tfrac{1}{2}\beta^2 V + \text{Var}(e_{ij}). \tag{6.9}$$

Hence

$$\text{Var}(a_i) = (\rho - \tfrac{1}{2}\beta^2)V$$
$$\text{Var}(e_{ij}) = (1 - \rho)V. \tag{6.10}$$

If there are $n_i$ children in the $i$th family with mean value $y_{i.}$, then

$$\text{Var}(y_{i.}|x_i) = \text{Var}(a_i) + \text{Var}(e_{ij})/n_i = \left[ (\rho - \tfrac{1}{2}\beta^2) + \frac{(1 - \rho)}{n_i} \right] V. \tag{6.11}$$

Thus we may write the regression of $y_{i.}$ on $x_i$ as

$$y_{i.} = m + \beta x_i + e_i \tag{6.12}$$

with $\text{Var}(e_i)$ given by eqn (6.11). If there are the same number of children in each family the regression can be estimated by ordinary least squares, but if the family sizes vary weighted least squares should be used. Define $w_i = 1/\text{Var}(e_i)$. The weighted estimate of $\beta$ is

$$\hat{\beta} = \sum w_i(x_i - \bar{x})(y_{i.} - \bar{y})/\sum w_i(x_i - \bar{x})^2 \tag{6.13}$$

with sampling variance

$$\text{Var}(\hat{\beta}) = 1/\sum w_i(x_i - \bar{x})^2. \tag{6.14}$$

To find the weights we may either estimate $\rho$ from the variance within sibships, as discussed in the next section, or we may assume that $\rho = \tfrac{1}{2}\beta$, in the absence of dominance and of environmental correlations between sibs. We must also have an estimate of $\beta$; an iterative procedure can be used starting with a trial value of $\beta$.

Let us now consider how many children should be reared and measured in each family in an experimental situation to estimate the regression as accurately as possible. We suppose that $N$ pairs of parents have been chosen at random, that each pair has $n$ children, and that $\beta = h^2$, $\rho = \frac{1}{2}h^2$. From eqn (6.14)

$$\text{Var}(\hat{\beta}) \simeq \frac{h^2\,(1-h^2)}{N} + \frac{(2-h^2)}{Nn}. \tag{6.15}$$

If the cost depends on the total number of children reared, then we want to minimize eqn (6.15) with $Nn$ fixed; this is done by minimizing $n$, that is to say by using as many parents as possible with one child each. The sampling variance with $n = 1$ is

$$\text{Var}(\hat{\beta}) \simeq (2-h^4)/N \simeq 2/N \text{ for small } h^2. \tag{6.16}$$

The number of families required to estimate the heritability to a desired accuracy can be found from this formula. On the other hand, it may be more realistic to equate the cost to the total number of individuals measured (including both children and parents), so that we want to minimize eqn (6.15) with $N(n + 2)$ fixed. This gives

$$n^2 = 2(2-h^2)/h^2(1-h^2). \tag{6.17}$$

For values of $h^2$ between $\frac{1}{4}$ and $\frac{3}{4}$ this gives an optimal family size of about four children.

Another design frequently used in agricultural experiments is to choose a number of sires at random and to mate each of them to several dams. Suppose for simplicity that each dam has one offspring (as would be the case in cattle) and that the phenotypic values of the sires are unknown; even if they are known little information is lost by ignoring them. The statistical model is

$$y_{ij} = m + a_i + \beta x_{ij} + e_{ij}. \tag{6.18}$$

In this model $x_{ij}$ is the value of the $j$th dam mated to the $i$th sire, $y_{ij}$ is the value of her child, $m$ is the overall mean, $a_i$ represents the effect of the $i$th sire, and $e_{ij}$ is the residual error. We now estimate the sire effects $(m + a_i)$ and the regression $\beta$ by least squares, which is equivalent to fitting a set of parallel lines; the estimate of $\beta$ is

$$\hat{\beta} = \sum (x_{ij} - x_{i.})(y_{ij} - y_{i.})/\sum (x_{ij} - x_{i.})^2 \tag{6.19}$$

with sampling variance

$$\text{Var}(\hat{\beta}) = \text{Var}(e_{ij})/\sum (x_{ij} - x_{i.})^2. \tag{6.20}$$

In the absence of epistasis and environmental correlation $\hat{\beta}$ estimates $\frac{1}{2}h^2$. $\text{Var}(e_{ij})$ can be estimated from the residual mean square. We can also find the theoretical value of the variances of $a_i$ and $e_{ij}$ in the same way as before; eqn

(6.10) must be modified by replacing $\frac{1}{2}\beta^2$ by $\beta^2$ and by interpreting $\rho$ as the correlation between half-sibs. If we write $\rho = \frac{1}{4}h^2$, we find that

$$\mathrm{Var}(\hat{\beta}) \simeq \frac{(1 - \frac{1}{4}h^2)}{(N - s)} \simeq 1/(N - s) \qquad (6.21)$$

where $N$ is the total number of offspring and $s$ is the number of sires. (Note that the Expected value of the denominator of eqn (6.20) is $(N - s)V$.) If we estimate $\hat{h}^2 = 2\hat{\beta}$, then $\mathrm{Var}\,(\hat{h}^2) \simeq 4/(N - s)$. This is the intra-sire regression of offspring on dam.

*Half-sib analysis*

Suppose that a group of sires is chosen at random, that each sire is mated to several dams, and that each dam produces one offspring. The phenotypic values of the offspring but not of the dams are measured. Thus offspring of the same sire are half-sibs, while offspring of different sires are unrelated. Heritability is estimated from the correlation between half-sibs. This experimental design is useful for uniparous animals such as cattle. It is often easier to obtain comparable measurements on groups of half-sibs belonging to the same generation than on parent and offspring belonging to different generations.

Suppose that there are $s$ sires, and that each of them has $n$ offspring by different dams. Among the offspring there are $s$ families, each consisting of $n$ half-sibs. Write $y_{ij}$ for the phenotypic value of the $j$th offspring of the $i$th sire. The data can be analysed by a one-way analysis of variance as shown in Table 6.3.

The model underlying this analysis is

$$y_{ij} = m + a_i + e_{ij}. \qquad (6.22)$$

In this equation $m$ is the overall mean, $a_i$ represents the effect of the $i$th sire and $e_{ij}$ is a residual error term representing variability within half-sib families. This is a random model, the $a_i$s being regarded as a random sample of sire effects drawn from a distribution with $E(a_i) = 0$. The $e_{ij}$s are also random variables, uncorrelated with $a_i$. Write $V$ for the phenotypic variance and $\rho$ for the correlation between half-sibs, so that

$$\mathrm{Var}(y_{ij}) = V$$
$$\mathrm{Cov}(y_{ij}, y_{ik}) = \rho V.$$

We also observe that

$$\mathrm{Var}(y_{ij}) = \mathrm{Var}(a_i) + \mathrm{Var}(e_{ij})$$
$$\mathrm{Cov}(y_{ij}, y_{ik}) = \mathrm{Var}(a_i).$$

It follows that

$$\mathrm{Var}(a_i) = \rho V$$
$$\mathrm{Var}(e_{ij}) = (1 - \rho)V. \qquad (6.23)$$

The Expected mean squares shown in Table 6.3 follow from these results.

TABLE 6.3

*Analysis of variance of half-sib design*

| Source of variation | Sum of squares | Degrees of freedom | Expected mean square |
|---|---|---|---|
| Between families | $n \sum (y_{i.} - y_{..})^2$ | $(s - 1)$ | $nV_a + V_e = n\rho V + (1 - \rho)V$ |
| Within families | $\sum (y_{ij} - y_{i.})^2$ | $s(n - 1)$ | $V_e = (1 - \rho)V$ |

We find from the Expected mean squares that $\rho$ can be estimated by

$$\hat{\rho} = \frac{(M_1 - M_2)}{M_1 + (n - 1)M_2} \qquad (6.24)$$

where $M_1$ and $M_2$ are the observed mean squares between and within families. This estimate is an intra-class correlation (Fisher 1921, 1946). If there is no epistasis and no environmental correlation between half-sibs, then $\rho = \frac{1}{4}h^2$; thus the heritability can be estimated by $\hat{h}^2 = 4\hat{\rho}$. It should be remembered, however, that both epistasis and environmental correlation will inflate this estimate of heritability.

If the $a_i$s and the $e_{ij}$s are normally distributed then the ratio

$$R \equiv \frac{E(M_2)M_1}{E(M_1)M_2} = \frac{(1 - \rho)}{1 + (n - 1)\rho} \frac{M_1}{M_2} \qquad (6.25)$$

will follow an $F$ distribution. A confidence interval for $\rho$ can be obtained by asserting that this quantity lies between the appropriate lower and upper percentage points of the $F$ distribution. If we assume that $h^2 = 4\rho$, this leads to a confidence interval for $h^2$.

It is also useful to have an approximate standard error for $\hat{\rho}$. To obtain this we use the so-called 'delta technique', which we will first explain in a more general context. Suppose that $X_1, X_2, \ldots, X_k$ is a set of random variables with known means, variances and covariances:

$$E(X_i) = \xi_i$$

$$\mathrm{Var}(X_i) = V_{ii}$$

$$\mathrm{Cov}(X_i, X_j) = V_{ij}.$$

We wish to find an approximate expression for the variance of some function of the $X_i$s, say $Y = f(\mathbf{X})$. To do this we expand $f(\mathbf{X})$ in a Taylor series about $\mathbf{X} = \boldsymbol{\xi}$:

$$Y = f(\boldsymbol{\xi}) + \sum_{i=1}^{k} (X_i - \xi_i) \frac{\partial f}{\partial X_i}\bigg|_{\mathbf{X} = \boldsymbol{\xi}}$$

+ higher terms which are ignored.

Hence

$$E(Y) \simeq f(\boldsymbol{\xi})$$

(6.26)

$$\text{Var}(Y) = E[Y - E(Y)]^2 \simeq \sum_{i,j} V_{ij} \frac{\partial f}{\partial X_i} \frac{\partial f}{\partial X_j}\bigg|_{\mathbf{X} = \boldsymbol{\xi}}.$$

This approximate expression for $\text{Var}(Y)$ is useful in many contexts. It can be generalized in an obvious way to find the covariance of two functions of the $X_i$s, say $Y_1 = f_1(\mathbf{X})$ and $Y_2 = f_2(\mathbf{X})$:

$$\text{Cov}(Y_1, Y_2) \simeq \sum_{i,j} V_{ij} \frac{\partial f_1}{\partial X_i} \frac{\partial f_2}{\partial X_j}\bigg|_{\mathbf{X} = \boldsymbol{\xi}}.$$

(6.27)

To apply this technique in the present case we first express $\hat{\rho}$ as a function of $R$, defined in eqn (6.25), and we use the known variance of $R$, which follows from the fact that it follows an $F$ distribution. We find that

$$\text{Var}(\hat{\rho}) = \text{Var}[f(R)] \simeq f'^2(R)|_{R=1} V(R) \simeq \frac{2(1-\rho)^2 [1 + (n-1)\rho]^2}{n(n-1)(s-1)}.$$

(6.28)

We can now ask, What is the most efficient design of experiment to minimize the sampling variance? Assume that the total number of animals to be measured is fixed at $sn = T$. If $n$ is large

$$\text{Var}(\hat{\rho}) \simeq 2(1-\rho)^2 (1 + n\rho)^2/nT.$$

This is minimized for fixed $T$ when $n = 1/\rho$. For example, if we are planning a half-sib experiment on a character whose heritability is expected to be about $\frac{1}{4}$, so that $\rho = \frac{1}{16}$, we should arrange to have about 16 offspring from each sire. In this example, $\text{Var}(\hat{\rho}) \simeq \frac{1}{2}T$, so that to obtain a standard error of, say, $\frac{1}{4}\rho = \frac{1}{64}$ we should need to use 128 sires each having 16 offspring.

We now consider the problem which arises when there are different numbers of individuals in different families. Suppose that there are $n_i$ half-sibs in the $i$th family. The analysis of variance corresponding to Table 6.3 is shown in Table 6.4. We can now estimate $V_a$ and $V_e$, and hence $\rho$, by equating observed and Expected sums of squares as before. It should be noted however that $M_1$ is no longer distributed like chi-square, and so $M_1/M_2$ is no longer distributed like $F$, since the individual terms in $M_1$ do not have the same variance unless $V_a = 0$. It also seems likely that this method of estimation will be inefficient when $V_a > 0$, since there is no logical reason why the term $(y_{i.} - y_{..})^2$ should receive the weight $n_i$ in estimating the variability between family means. In particular, if $V_a \gg V_e$ the family means will be equally variable whether they are from small or large families and should thus receive equal weight.

TABLE 6.4

*Analysis of variance of half-sibs design with unequal numbers*

| Source of variation | Sum of squares | Degrees of freedom | Expected mean square |
|---|---|---|---|
| Between families | $S_B = \sum n_i (y_{i.} - y_{..})^2$ | $(s-1)$ | $\dfrac{(N^2 - \sum n_i^2)}{N(s-1)} V_a + V_e$ |
| Within families | $S_W = \sum (y_{ij} - y_{i.})^2$ | $N-s$ | $V_e$ |

Note that $y_{..} = \sum n_i y_{i.}/N,\, N = \sum n_i$.

By analogy with weighted regression we are led to consider the weighted sum of squares between families

$$S_B^* = \sum w_i (y_{i.} - y_{..})^2 \tag{6.29}$$

where

$$w_i = 1/\mathrm{Var}(y_{i.}) = (V_a + V_e/n_i)^{-1}$$

$$y_{..} = \sum w_i y_{i.}/\sum w_i.$$

The best estimate of the within family variance, $V_e$, is obviously obtained from the within families sum of squares, $S_w$, defined in Table 6.4. $S_B^*$ has a chi-square distribution with $(s-1)$ degrees of freedom and is distributed independently of $S_W/V_e$ which is also chi-square. We therefore define

$$F = \frac{S_B^*}{(s-1)} \div \frac{S_W}{(N-s)V_e} = \frac{(N-s)}{(s-1)} \frac{\sum \left(\dfrac{V_a}{V_e} + \dfrac{1}{n_i}\right)^{-1} (y_{i.} - y_{..})^2}{\sum (y_{ij} - y_{i.})^2}. \tag{6.30}$$

This quantity is distributed as an $F$-ratio with $(s-1)$ and $(N-s)$ degrees of freedom. By putting $F = 1$ we obtain an implicit equation for estimating $V_a/V_e$, and hence $\rho$; this equation can be solved iteratively. (Note that $y_{..}$ also depends on $V_a/V_e$ and should therefore be recalculated in each iteration as well as the weights.) A confidence interval can be obtained by equating $F$ to the appropriate upper and lower percentage points of the $F$ distribution.

*Full-sib analysis*

For multiparous animals, such as pigs, which have several young in a litter, a simple experimental design is to mate each sire with one dam and to measure $n$ offspring from each litter. The members of each family are full sibs, and the same analysis can be used as before if $\rho$ is interpreted as the correlation between full sibs. The heritability can be estimated as $2\rho$ if it can be assumed that there is no epistasis, no dominance and no environmental correlation.

However, these assumptions are rather unrealistic; in particular there may well be an environmental correlation between litter mates.

To overcome this problem it is better to mate each sire with several dams, so that both full sibs and half sibs are obtained. Suppose that there are $s$ sires, that each sire is mated to $d$ dams and that $n$ offspring are measured from each litter. Members of the same litter are full sibs, members of different litters with different sires are unrelated. Each dam has only one litter, so that there are $sd$ dams in all.

This is a two-way nested design, and the appropriate analysis is shown in Table 6.5. The model underlying the analysis is

$$y_{ijk} = m + a_i + b_{ij} + e_{ijk}. \tag{6.31}$$

In this equation $y_{ijk}$ is the value of the $k$th offspring of the $j$th dam of the $i$th sire, $m$ is the overall mean, $a_i$ represents the effect of the $i$th sire, $b_{ij}$ the effect of the $j$th dam mated to this sire, and $e_{ijk}$ is a residual error term representing variability within full sib families.

TABLE 6.5

*Analysis of variance of full-sib design*

| Source of variation | Sum of squares | Degrees of freedom | Expected mean square |
|---|---|---|---|
| Between sires | $nd \sum (y_{i..} - y_{...})^2$ | $(s-1)$ | $nd\rho_H V + n(\rho_F - \rho_H)V + (1-\rho_F)V$ |
| Between dams within sires | $n \sum (y_{ij.} - y_{i..})^2$ | $s(d-1)$ | $n(\rho_F - \rho_H)V + (1-\rho_F)V$ |
| Within litters | $\sum (y_{ijk} - y_{ij.})^2$ | $sd(n-1)$ | $(1-\rho_F)V$ |

Denote the correlations between full and half sibs by $\rho_F$ and $\rho_H$ respectively. Then

$$\mathrm{Var}(y_{ijk}) = \mathrm{Var}(a_i) + \mathrm{Var}(b_{ij}) + \mathrm{Var}(e_{ijk}) = V$$

$$\mathrm{Cov}(y_{ijk}, y_{ijk'}) = \mathrm{Var}(a_i) + \mathrm{Var}(b_{ij}) = \rho_F V \tag{6.32}$$

$$\mathrm{Cov}(y_{ijk}, y_{ij'k'}) = \mathrm{Var}(a_i) = \rho_H V.$$

Thus

$$\mathrm{Var}(a_i) = \rho_H V$$

$$\mathrm{Var}(b_{ij}) = (\rho_F - \rho_H)V \tag{6.33}$$

$$\mathrm{Var}(e_{ijk}) = (1 - \rho_F)V.$$

The expected mean squares shown in Table 6.5 follow from these results. The obvious estimates of $\rho_F$, $\rho_H$, and $V$ are obtained by equating the observed mean squares to their Expected values; the heritability is then estimated as $4 \hat{\rho}_H$.

TABLE 6.6

*Sib analysis of body length in pigs*

| Source of variation | Degrees of freedom | Mean square | Expected mean square |
|---|---|---|---|
| Between sires | 432 | 6.03 | $(1 + \rho_F + 2\rho_H)V$ |
| Between dams within sires | 468 | 3.81 | $(1 + \rho_F - 2\rho_H)V$ |
| Within litters | 936 | 2.87 | $(1 - \rho_F)V$ |

To illustrate this procedure we consider the data of Fredeen and Jonsson (1957), quoted by Falconer (1960), on the inheritance of body length in pigs. 468 sires were each mated to two dams, and the length of two male offspring in each litter was recorded. The analysis of variance is shown in Table 6.6; 36 degrees of freedom have been lost from the between sires sum of squares in eliminating differences between testing stations and between years, but this makes no difference to the Expected mean squares. Equating observed and Expected mean squares, we find that

$$\hat{\rho}_H = (M_1 - M_2)/(M_1 + M_2 + 2M_3) = 0.142$$

$$\hat{\rho}_F = (M_1 + M_2 - 2M_3)/(M_1 + M_2 + 2M_3) = 0.263,$$

where $M_i$ is the $i$th mean square.

To find approximate standard errors for these estimates we note that, on the assumption of normality, the mean squares $M_i$ with $f_i$ degrees of freedom are independently distributed with variances which can be estimated by $\text{Var}(M_i) = 2M_i^2/(f_i + 2)$. (See Chapter 5.) Since $\hat{\rho}$ is the ratio of two linear functions of the mean squares, we can find its sampling variance from the 'delta technique' formula for the approximate variance of a ratio

$$\text{Var}(X_1/X_2) = \left(\frac{\xi_1}{\xi_2}\right)^2 \left\{ \frac{\text{Var}(X_1)}{\xi_1^2} + \frac{\text{Var}(X_2)}{\xi_2^2} - \frac{2\,\text{Cov}(X_1, X_2)}{\xi_1\,\xi_2} \right\}$$

which is easily derived from eqn (6.26). After substituting observed for Expected values of $X_1$ and $X_2$, we find that

$$\text{Var}(\hat{\rho}_H) = 0.00085$$

$$\text{Var}(\hat{\rho}_F) = 0.00099$$

$$\text{Cov}(\hat{\rho}_H, \hat{\rho}_F) = 0.00028.$$

(The formula for the covariance of two ratios is easily derived from eqn (6.27).)

The estimate of heritability obtained by multiplying $\hat{\rho}_H$ by 4 is $0.57 \pm 0.12$; the estimate obtained by doubling $\hat{\rho}_F$ is $0.53 \pm 0.06$. There is in this case no evidence of any difference between these estimates. If we assume that they are both estimates of the heritability, we may seek as its best estimate their

weighted average with minimum variance which is 0.54 $\pm$ 0.06, giving weights
of 0.14 and 0.86 to the two estimates.

## *Pedigree analysis

An alternative approach which has been developed recently is to write down
the likelihood of the observations and to estimate the variance components by
maximum likelihood. Suppose that observations have been made on a number
of individuals in a single pedigree. Write **R** for the matrix of $R$ values and **P**$_2$ for
the matrix of $P_2$ values of the pairs of individuals, who may be related in any
way which excludes inbreeding. In the absence of epistasis and of environmental correlations, the variance–covariance matrix of the observations is

$$\mathbf{V} = V_A \mathbf{R} + V_D \mathbf{P}_2 + V_E \mathbf{I}.$$

We also assume that the observations are multivariate normal with the same
mean $m$. (See page 123 for discussion of the assumptions underlying multivariate normality.) The logarithm of the likelihood is

$$L = \text{constant} - \tfrac{1}{2} \ln|\mathbf{V}| - \tfrac{1}{2}(\mathbf{y} - m\mathbf{1})^T \mathbf{V}^{-1} (\mathbf{y} - m\mathbf{1}),$$

where **y** is the vector of the observations and **1** is a column of *1s*. In practice a
large number of unrelated pedigrees will have been observed; their total log
likelihood is obtained by summing over the individual pedigrees.

The maximum likelihood scores for the parameters $m$, $V_A$, $V_D$, and $V_E$ are

$$\partial L/\partial m = \mathbf{1}^T \mathbf{V}^{-1} (\mathbf{y} - m\mathbf{1})$$

$$\partial L/\partial V_\alpha = -\tfrac{1}{2}tr\left(\mathbf{V}^{-1} \frac{\partial \mathbf{V}}{\partial V_\alpha}\right) + \tfrac{1}{2}(\mathbf{y} - m\mathbf{1})^T \mathbf{V}^{-1} \frac{\partial \mathbf{V}}{\partial V_\alpha} \mathbf{V}^{-1} (\mathbf{y} - m\mathbf{1}),$$

where $\partial \mathbf{V}/\partial V_\alpha = \mathbf{R}$, $\mathbf{P}_2$, or $\mathbf{I}$ according as $V_\alpha = V_A$, $V_D$, or $V_E$. The elements of
the information matrix are

$$-E(\partial^2 L/\partial m^2) = \mathbf{1}^T \mathbf{V}^{-1} \mathbf{1}$$

$$-E(\partial^2 L/\partial m \, \partial V_\alpha) = 0$$

$$-E(\partial^2 L/\partial V_\alpha \, \partial V_\beta) = \tfrac{1}{2}tr\left(\mathbf{V}^{-1} \frac{\partial \mathbf{V}}{\partial V_\alpha} \mathbf{V}^{-1} \frac{\partial \mathbf{V}}{\partial V_\beta}\right).$$

The maximum likelihood estimates can be found iteratively by equating the
scores to zero (after adding over pedigrees). The inverse of the information
matrix (after summing the latter over pedigrees) is used in the iterative estimation procedure, and also gives the variances and covariances of the
estimates. For an account of the method of maximum likelihood see, for
example, Bailey (1961), Appendix 1); for a review of its application to pedigree

analysis see Boyle and Elston (1979). In applying the method to the estimation of variance components, some authors advocate a modification called restricted maximum likelihood; see Harville (1977) for a review of this problem.

The advantage of this method is its flexibility. It can be applied to pedigrees of any kind, and the model can be extended to allow for mean values which depend on age or sex or on environmental variables and to incorporate further variance components, such as a subdivision of the environmental variance into components between and within mothers to allow for maternal effects.

### *Epistasis

It has been assumed so far that there are no epistatic interactions between loci. In the presence of epistasis but without linkage the genetic covariance between relatives is given by

$$\text{Cov}(G', G'') = RV_A + P_2 V_D + R^2 V_{AA} + RP_2 V_{AD} + P_2^2 V_{DD}$$
$$+ R^3 V_{AAA} + \cdots \qquad (6.34)$$

To illustrate the probable occurrence of epistasis, Table 6.7 shows the correlations between relatives for diversity of fingerprint ridge count, calculated as the standard deviation of the measurements from the ten fingers (Holt 1968); this table should be compared with the results for total ridge count in Table 6.2. The regression of child on mid-parent is exactly twice the correlation between sibs, which suggests that there is neither dominance nor environmental correlation, since both these factors would be expected to inflate the correlation between sibs. However, the correlation between monozygotic twins is much higher than the regression of child on mid-parent, which suggests a component

TABLE 6.7

*Correlations between relatives for diversity of fingerprint ridge count (Holt 1968)*

| Relationship | Observed value ± standard error | Theoretical value Coefficient of | | | | | |
|---|---|---|---|---|---|---|---|
| | | $\dfrac{V_A}{V_Y}$ | $\dfrac{V_D}{V_Y}$ | $\dfrac{V_{AA}}{V_Y}$ | $\dfrac{V_{AD}}{V_Y}$ | $\dfrac{V_{DD}}{V_Y}$ | $\dfrac{V_{AAA}}{V_Y}$ $\cdots$ |
| Monozygotic twins | $0.73 \pm 0.06$ | $1$ | $1$ | $1$ | $1$ | $1$ | $1$ $\cdots$ |
| Dizygotic twins | $0.22 \pm 0.10$ | $\frac{1}{2}$ | $\frac{1}{4}$ | $\frac{1}{4}$ | $\frac{1}{8}$ | $\frac{1}{16}$ | $\frac{1}{8}$ $\cdots$ |
| Sibs | $0.22 \pm 0.03$ | $\frac{1}{2}$ | $\frac{1}{4}$ | $\frac{1}{4}$ | $\frac{1}{8}$ | $\frac{1}{16}$ | $\frac{1}{8}$ $\cdots$ |
| Child on mid-parent regression | $0.44 \pm 0.06$ | $1$ | $0$ | $\frac{1}{2}$ | $0$ | $0$ | $\frac{1}{4}$ $\cdots$ |

of variance due to epistasis. If we assume that there are only first order interactions we obtain the percentage estimates $V_A = 15$ per cent, $V_{AA} = 58$ per cent, $V_E = 27$ per cent, but the data are consistent with many other possibilities if higher order interactions are introduced; information about other types of relatives would be needed to distinguish between them. It will be noted that no departures from completely additive gene action would have been suspected without the evidence from identical twins.

We shall now derive eqn (6.34). The derivation is rather complicated and the reader may wish to postpone detailed study until a second reading. We first introduce some more notation. Given a set $S$ marking the positions of some of the genes in the genotype (see page 46), we define $x_i$ as the number of genes at the $i$th locus contained in $S$ and $\mathbf{x}(S)$ as the vector of the $x_i$s; formally, $x_i = \chi(i) + \chi(n + i)$, where $\chi$ is the indicator function of the set $S$, $\chi(i) = 1$ or $0$ according as $i$ does or does not belong to $S$. Two sets $S$ and $T$ will be said to be similar if $\mathbf{x}(S) = \mathbf{x}(T)$.

Suppose now that $S$ and $T$ refer respectively to the positions of genes in two relatives. We define $y_i$ as the number of pairs of genes which are identical by descent among the genes at the $i$th locus in $S$ and $T$; the corresponding vector is $\mathbf{y}(S, T)$. We may imagine that the set $T$ is enumerated below $S$ and that identical genes are joined by lines as in Fig. 3.1; note that identical genes must occupy the same locus. It is clear that $y_i$ cannot exceed the smaller of $x_i(S)$ and $x_i(T)$. If $S$ and $T$ are similar so that $\mathbf{x}(S) = \mathbf{x}(T)$, and if furthermore $\mathbf{y}(S, T) = \mathbf{x}(S)$, then for each gene in $S$ there corresponds an identical gene in $T$ and vice versa. In this case we shall say that the relatives are identical with respect to the ordered pair of sets $S$, $T$.

To find the genetic covariance in the general case we first decompose the genotypic values of the two relatives under random mating:

$$G' = m + \sum_{\emptyset \subset S \subseteq N} X'_S$$

$$G'' = m + \sum_{\emptyset \subset S \subseteq N} X''_S.$$

We shall now evaluate the covariance of a typical product $X'_S X''_T$. If the relatives are identical with respect to $S$, $T$, then $X'_S = X''_T$ so that

$$E(X'_S X''_T) = \mathrm{Var}(X_S).$$

If the relatives are not identical with respect to $S$, $T$ then

$$E(X'_S X''_T) = 0.$$

For in this case there must be a gene in $S$ or $T$ which is not identical with any other gene in $S$ or $T$ and which is therefore statistically independent of all these other genes. The result follows by averaging over this gene.

Consider now the interaction between additive effects at $a$ specified loci and the dominance deviations at $d$ other specified loci; this interaction is the sum of

$2^a$ components $X_S$ as represented in eqn (4.6). To obtain a general way of writing this interaction, let $\mathbf{z}$ be a vector of $n$ elements whose $k$th element $z_k$ shows how the $k$th locus occurs in the interaction: $z_k = 0$ means that the locus does not occur in the interaction, $z_k = 1$ that its additive effect occurs, $z_k = 2$ that its dominance deviation occurs. The interaction can be written

$$I_{\mathbf{z}} = \Sigma X_S,$$

the sum being over all sets $S$ for which $\mathbf{x}(S) = \mathbf{z}$. If $S$ is any one of these sets, then

$$\mathrm{Var}(I_{\mathbf{z}}) = 2^a \, \mathrm{Var}(X_S),$$

since each $X_S$ in the sum has the same variance.

Different interactions in the two relatives are uncorrelated; that is to say, if $\mathbf{z} \neq \mathbf{z}^*$, then

$$E(I'_{\mathbf{z}} \, I''_{\mathbf{z}^*}) = 0.$$

The reason is that any pair $X'_S$, $X''_T$ occurring in the expansion of $I'_{\mathbf{z}} \, I''_{\mathbf{z}^*}$ must be uncorrelated since $S$ and $T$ cannot be similar so that the relatives cannot be identical with respect to $S$, $T$. The covariance between the same interactions in the two relatives is

$$
\begin{aligned}
E(I'_{\mathbf{z}} \, I''_{\mathbf{z}}) &= \Sigma \, Pr[\mathbf{y}(S, T) = \mathbf{z}] \, \mathrm{Var}(X_S) \\
&= 2^{-a} \, \mathrm{Var}(I_{\mathbf{z}}) \, \Sigma \, Pr[\mathbf{y}(S, T) = \mathbf{z}],
\end{aligned}
\tag{6.35}
$$

the sum being over all sets $S$ and $T$ with $\mathbf{x}(S) = \mathbf{x}(T) = \mathbf{z}$.

To evaluate this sum we shall assume that all the loci are unlinked so that the numbers of pairs of identical genes at different loci are statistically independent. Thus

$$Pr[\mathbf{y}(S, T) = \mathbf{z}] = \prod_{i=1}^{n} Pr(y_i = z_i).$$

If $z_i = 2$, then $Pr(y_i = 2) = P_2$; furthermore the pair $S$, $T$ is invariant at the positions representing the $i$th locus. If $z_i = 1$, the pair $S$, $T$ has four variations at the $i$th locus depending on whether the gene represented is of maternal or paternal origin in each of the two relatives. Reference to Fig. 3.1 will show that the sum of the probabilities that $y_i = 1$ over these four variations is $(P_1 + 2P_2)$ $= 2R$. If $z_i = 0$, then $Pr(y_i = 0) = 1$ trivially. It follows that

$$Pr[\mathbf{y}(S, T) = \mathbf{z}] = (2R)^a \, P_2^d.
\tag{6.36}$$

Hence

$$\mathrm{Cov}(I'_{\mathbf{z}}, I''_{\mathbf{z}}) = R^a \, P_2^d \, \mathrm{Var}(I_{\mathbf{z}}).
\tag{6.37}$$

We now write $I$ for the total interaction between $a$ additive effects and $d$ dominance deviations, so that $I = \sum I_z$, the sum being over all $z$ with $a$ 1s and $d$ 2s. Then

$$\text{Cov}(I', I'') = R^a P_2^d \text{Var}(I) \tag{6.38}$$

Decomposing the genotypic value in terms of these interactions as in eqn (4.7) we obtain the final result (6.34).

### Linkage

It was assumed in obtaining eqn (6.34) that all loci are unlinked so that the numbers of pairs of identical genes at different loci are statistically independent. The result remains valid for parent and child in the presence of linkage because they always share exactly one pair of identical genes at each locus. For other relatives the coefficients of $V_A$ and $V_D$ remain unchanged by linkage, because the components of $A$ and $D$ only depend on the genes at a single locus, but the coefficients of the interaction variances will be increased by linkage.

Consider two loci and write $P(i, j)$ for the probability that there are $i$ pairs of identical genes at the first locus and $j$ pairs at the second. By using eqn (6.34) in conjunction with Fig. 3.1 we find the following results for the coefficients of the interaction variances in the genetic covariance:

| $z$ | $I_z$ | Coefficient of $\text{Var}(I_z)$ |
|-----|-------|----------------------------------|
| (1, 1) | $AA_{12}$ | $\frac{1}{4}P(1, 1) + \frac{1}{2}P(1, 2) + \frac{1}{2}P(2, 1) + P(2, 2)$ |
| (1, 2) | $AD_{12}$ | $\frac{1}{2}P(1, 2) + P(2, 2)$ |
| (2, 1) | $AD_{21}$ | $\frac{1}{2}P(2, 1) + P(2, 2)$ |
| (2, 2) | $DD_{12}$ | $P(2, 2)$ |

The general result for an arbitrary number of loci is obvious by inspection.

To obtain some idea of the magnitude of this effect, consider the covariance between sibs under a two locus model. The calculation of $P(i, j)$ was obtained in Problem 3.9. The covariance is found to be

$$\text{Cov}(G', G'') = V_A/2 + V_D/4 + (2 + c)V_{AA}/8 + (1 + c)V_{AD}/8$$
$$+ (1 + c)^2 V_{DD}/16$$

where $c = (1 - 2r)^2$, $r$ being the recombination fraction. Some numerical results are given in Table 6.8.

### Correlated and sex-linked characters

It is a commonplace of Mendelian genetics that genes are pleiotropic in their action; in other words, one gene may affect many different characters. It is to be expected that genes affecting quantitative characters will also be pleiotropic, which will lead to genetic correlations between these characters.

TABLE 6.8

*Genetic covariances between sibs arising from two linked loci*

| Recombination fraction | $V_A$ | $V_D$ | $V_{AA}$ | $V_{AD}$ | $V_{DD}$ |
|---|---|---|---|---|---|
| 0.5 | 0.5 | 0.25 | 0.25 | 0.125 | 0.0625 |
| 0.25 | 0.5 | 0.25 | 0.28 | 0.156 | 0.0977 |
| 0.1 | 0.5 | 0.25 | 0.33 | 0.205 | 0.1681 |
| 0.05 | 0.5 | 0.25 | 0.35 | 0.226 | 0.2048 |
| 0 | 0.5 | 0.25 | 0.37 | 0.250 | 0.2500 |

Consider two characters with phenotypic values $Y_1$ and $Y_2$ which can be decomposed into genetic and environmental components in the usual way:

$$Y_1 = G_1 + E_1$$
$$Y_2 = G_2 + E_2.$$

We observe that $E_1$ and $E_2$ are uncorrelated with, but not necessarily independent of, $G_1$ and $G_2$ since genotypes and environments are assumed to be independent (see page 20). Hence

$$\text{Cov}(Y_1, Y_2) = \text{Cov}(G_1, G_2) + \text{Cov}(E_1, E_2). \tag{6.39}$$

Thus a phenotypic correlation may be due either to a genetic correlation (pleiotropy) or an environmental correlation.

We now decompose the genotypic values in the usual way:

$$G_1 = m_1 + A_1 + D_1 + AA_1 + \cdots$$
$$G_2 = m_2 + A_2 + D_2 + AA_2 + \cdots,$$

$A_1$ being the total additive effect of the first character, and so on. Then

$$\text{Cov}(G_1, G_2) = \text{Cov}(A_1, A_2) + \text{Cov}(D_1, D_2) + \text{Cov}(AA_1, AA_2) + \cdots \tag{6.40}$$

Note that cross-terms, such as $\text{Cov}(A_1, D_2)$, are zero; this can be proved by a simple extension of the result (4.14), which shows that terms like $\text{Cov}(A_1, D_1)$ are zero. These covariances refer to covariances between the components of the genotypic values of the two characters in the same individual. Information about them can be obtained by measuring the correlation between the phenotypic values of the characters in relatives. Let $G_1'$ and $G_2''$ be the genotypic values of the two characters in related individuals. If we assume autosomal inheritance an obvious extension of the argument leading to eqn (6.38) shows that

$$\text{Cov}(G_1', G_2'') = R\,\text{Cov}(A_1, A_2) + P_2\,\text{Cov}(D_1, D_2) + R^2\,\text{Cov}(AA_1, AA_2) + \cdots \tag{6.41}$$

Genetic correlations can be estimated from parent–offspring regressions or from data on sibs or half sibs. To illustrate the former situation, consider some data given by Hazel (1943) on two characters in pigs: $Y_1$ = weight at 180 days, $Y_2$ = a numerical score of market suitability assessed by a panel of judges. The phenotypic variances and the covariance are shown at the top of Table 6.9. The regression coefficients of offspring on dam are

$$b_{11} = 0.149, \quad b_{12} = 0.153, \quad b_{21} = 0.013, \quad b_{22} = 0.049$$

where $b_{ij}$ is the regression of trait $i$ in the offspring on trait $j$ in the dam. The genetic variances for weight and score are estimated as $2b_{11}V_{11}$ and $2b_{22}V_{22}$ respectively, where $V_{ij}$ represents a phenotypic variance (or covariance). The genetic covariance can be estimated in two ways, as $2b_{12}V_{22}$ or as $2b_{21}V_{11}$, and it seems natural to use their average, $b_{12}V_{22} + b_{21}V_{11}$ as a combined estimate. The environmental variances and covariance are obtained by subtraction. This method provides a valid estimate of the additive genetic covariance in the absence of epistasis; the environmental component would also include any contribution from dominance. It will be seen that the estimated correlations are quite high.

TABLE 6.9

*Estimates of phenotypic and genetic variances and covariances for two traits in pigs (after Hazel 1943)*

|  |  | Weight | Score | Correlation |
|---|---|---|---|---|
| Phenotypic | Weight | 1015 | 94 | 0.62 |
|  | Score |  | 23 |  |
| Genetic | Weight | 302 | 17 | 0.69 |
|  | Score |  | 2 |  |
| Environmental | Weight | 713 | 77 | 0.63 |
|  | Score |  | 21 |  |

In general, write $y_{1i}$ and $y_{2i}$ for the values of the two characters in the $i$th parent (or mid-parent), and $y_{3i}$ and $y_{4i}$ for the corresponding mean values in the offspring of that parent, assuming for simplicity that each family has the same number of offspring. Calculate the mean squares and cross-products defined as

$$M_{jk} = \sum_i (y_{ji} - y_{j.})(y_{ki} - y_{k.})/f, \quad j, k = 1, \ldots, 4, \tag{6.42}$$

where $f$ is the number of degrees of freedom. The genetic and environmental variances, covariances, and correlations can be estimated in terms of these quantities. In particular, the estimate of the genetic correlation is

$$r_G = \frac{M_{14} + M_{23}}{2(M_{13}M_{24})^{1/2}}. \tag{6.43}$$

The standard errors of these estimates can be found by the delta technique if we know the variances and covariances of the $M_{jk}$s. Assume that the observations are a random sample from a multinormal distribution and write $m_{jk} = E(M_{jk})$. By standard theory (Anderson 1958, p. 161)

$$\text{Cov}\,(M_{jk}, M_{lm}) = (m_{jl}\,m_{km} + m_{jm}\,m_{kl})/f. \tag{6.44}$$

If $j = l, k = m$ we obtain the variance

$$\text{Var}(M_{jk}) = (m_{jk}^2 + m_{jj}m_{kk})/f. \tag{6.45}$$

Unbiased estimators of these variances and covariances can be obtained by substituting $M_{jk}$ for $m_{jk}$, and dividing by $(f + 2)$ instead of $f$.

Genetic covariances can also be estimated from data on sibs or half sibs by doing a multivariate analysis of variance, which is the obvious extension of the univariate analysis in Tables 6.3 and 6.5. The genetic and environmental variances can be estimated from the analyses of variance on the two characters by equating observed and Expected values, if the Expected values are expressed in terms of components of variance; the covariances can be estimated from the corresponding analysis of covariance in which sums of cross-products replace sums of squares.

### TABLE 6.10

*Estimation of genetic variances and covariances for plant height and ear height in maize (Mode and Robinson 1959)*

(a) *Multivariate analysis of variance*

| Source of variation | Degrees of freedom | Mean square or cross-product | | | Expected value |
| --- | --- | --- | --- | --- | --- |
| | | 11 | 12 | 22 | |
| Between males | 48 | 77.6 | 35.5 | 38.6 | $8V_a + 2V_b + V_e$ |
| Between females (within males) | 192 | 30.7 | 15.0 | 11.4 | $2V_b + V_e$ |
| Within females | 239 | 10.0 | 4.1 | 4.0 | $V_e$ |

NB The mean squares for the two characters are shown in the columns headed 11 and 22, the mean cross-product in 12. $V_a$ represents $\text{Var}(a_1)$, $\text{Cov}(a_1, a_2)$, or $\text{Var}(a_2)$ corresponding to Expected values of the mean squares or products in columns 11, 12, or 22, and likewise for $V_b$ and $V_e$.

(b) *Estimates of components of variance and covariance with standard errors*

| | | Plant height | Ear height | Correlation |
| --- | --- | --- | --- | --- |
| Additive genetic | Plant height | $23.4 \pm 7.9$ | $10.2 \pm 4.7$ | $0.57 \pm 0.14$ |
| | Ear height | | $13.6 \pm 3.9$ | |
| Dominance | Plant height | $18.0 \pm 11.1$ | $11.2 \pm 6.4$ | Unreliable |
| | Ear height | | $1.2 \pm 4.9$ | |
| Environmental | Plant height | $7.5 \pm 1.2$ | $2.7 \pm 0.6$ | $0.55 \pm 0.07$ |
| | Ear height | | $3.2 \pm 0.5$ | |

To illustrate the method Table 6.10 shows the analysis of the correlations between plant height and ear height in maize (Mode and Robinson 1959). An $F_2$ generation was raised following a cross between two pure lines. (In the absence of linkage this generation is in Hardy–Weinberg and linkage equilibrium, so that methods designed for outbred populations are appropriate.) In the experiment four males from this $F_2$ were chosen at random, and each male was mated to four different females; offspring from each of these 16 matings were planted in two blocks, with 16 plots in each block and with each mating represented once in each block. Each plot contained about ten plants from a single mating, and the plot mean was the unit of observation. This design gives rise to an analysis of variance like Table 6.5, with three degrees of freedom between males, 12 between females within males, and 15 within females. (In Table 6.5 there would be 16 degrees of freedom within females, but in the present analysis one of them represents differences between blocks.) This experiment was then replicated 16 times, with different males and females, giving rise to the analysis of variance and covariance in Table 6.10a. (There was one missing plot, reducing the degrees of freedom within females by one.)

The model underlying the analysis is the bivariate analogue of eqn (6.31):

$$y_{1ijk} = m_1 + a_{1i} + b_{1ij} + e_{1ijk}$$
$$y_{2ijk} = m_2 + a_{2i} + b_{2ij} + e_{2ijk} \tag{6.46}$$

where $y_1$ and $y_2$ denote the two characters; a full model would also contain terms for blocks and replicates, but these do not concern us here and have been omitted for simplicity. The Expected values of the mean squares and products are shown in Table 6.10a in terms of the variances and covariances of the variables in eqn (6.46). These variances and covariances are easily estimated by equating observed and Expected values. In the absence of epistasis they can be expressed in terms of the genetic components of variance and covariance as follows:

$$V_a = 0.25V_A$$
$$V_b = 0.25V_A + 0.25V_D \tag{6.47}$$
$$V_e = V_{Eb} + 0.1V_{Ew} + 0.05V_A + 0.075V_D.$$

These results follow from eqn (6.33). As in the table, they express results for the variance of either of the characters or for the covariance between them. The third equation assumes that the environmental variance has two components, the variance between plots, $V_{Eb}$, and the variance between plants in the same plot, $V_{Ew}$. $V_e$ is the variance of the mean of ten full sibs in the same plot. If we define $V_E = V_{Eb} + 0.1V_{Ew}$, the genetic and environmental components of variance can be estimated from eqn (6.47), and the sampling variances and covariances of these estimates can be found from eqn (6.45), since the estimates are linear functions of the mean squares and products. Finally, the correlations can be estimated, and their sampling variances found by the delta technique. The results of these calculations are shown in Table 6.10b.

Correlated characters are of great importance in animal and plant breeding since selection for one character may bring about undesired responses in other characters which are correlated with it. We shall return to this question in Chapter 11. Another application is in the analysis of genotype–environment interaction in an outbred population. Suppose that $s$ sires are chosen at random and that each of them is mated to $nt$ dams, each dam having one offspring. The $nt$ offspring in each half-sib family are randomly divided into $t$ groups with $n$ individuals in each group, and the groups are then raised in $t$ different environments. An analysis of variance of the results is as follows:

| Source of variation | Sum of squares | Degrees of freedom | Expected mean square |
|---|---|---|---|
| Families | $nt \sum_{i} (y_{i..} - y_{...})^2$ | $(s-1)$ | $ntV_a + nV_I + V_e$ |
| Environments | $ns \sum_{j} (y_{.j.} - y_{...})^2$ | $(t-1)$ | $nsV_t + nV_I + V_e$ |
| Families × environments | $n \sum_{i,j} (y_{ij.} - y_{i..} - y_{.j.} + y_{...})^2$ | $(s-1)(t-1)$ | $nV_I + V_e$ |
| Within families and environments | $\sum_{i,j,k} (y_{ijk} - y_{ij.})^2$ | $(n-1)st$ | $V_e$ |

The model underlying this analysis is

$$y_{ijk} = m + a_i + t_j + I_{ij} + e_{ijk} \qquad (6.48)$$

where $m$ is the overall mean, $a_i$ is a family (sire) effect, $t_j$ is an environmental effect, and $I_{ij}$ the interaction between them. The Expected mean squares are given on the assumption that both family and environmental effects are random; if environments are fixed the Expected mean square for families loses the term in $V_I$. The interaction can be tested in the usual way, and the components of variance can be estimated by equating observed and Expected values.

If a significant interaction is found, its biological significance must be interpreted. To do this it is instructive, following Falconer (1952), to regard the phenotypic values observed in different environments as different, correlated characters. Writing $m_j = m + t_j$, $a_{ij} = a_i + I_{ij}$, the model (6.48) can be re-written

$$y_{ijk} = m_j + a_{ij} + e_{ijk}. \qquad (6.49)$$

In this reformulation $m_j$ is the mean value in the $j$th environment, and $a_{ij}$ is the sire effect of the $i$th sire in the $j$th environment, which is equal to half the breeding value in this environment. For a fixed environment, or pair of environments,

$$\text{Var}(a_{ij}) = \tfrac{1}{4}\text{Var}(A_j)$$
$$\text{Cov}(a_{ij}, a_{ij'}) = \tfrac{1}{4}\text{Cov}(A_j, A_{j'})$$

(6.50)

where $A_j$ is the breeding value in the $j$th environment. (We assume that there is no epistasis and no environmental correlation between half sibs.) The variances of sire effects in different environments can be estimated from the analyses of variance between and within families, holding the environment fixed. (See Table 6.3.) The covariance between sire effects in two environments can be estimated as

$$\sum (y_{ij.} - y_{.j.})(y_{ij'.} - y_{.j'.})/(s - 1).$$

(6.51)

Hence the correlation between sire effects can be calculated, which is also the correlation between breeding values (see eqn (6.50)). Note that this situation differs from the multivariate analysis of variance considered in Table 6.10 since in the present case each individual can only be observed in one environment.

These ideas can be extended to the situation in which the genotype reacts differently in males and females (genotype × sex interaction) by regarding the phenotypic values in the two sexes as different correlated characters. Care must be taken to distinguish genotype–sex interaction from sex-linkage, the effects of which we shall finally briefly consider.

### *Sex linkage

As remarked in Chapter 4 the genotypic values must be decomposed separately in males and females under sex-linkage since females are diploid but males are haploid at sex-linked loci. For simplicity we shall consider a character determined by a single sex-linked locus but the results will be of general use since in the absence of epistasis covariances from different loci, sex-linked or autosomal, are additive.

We shall first consider whether one can predict any relationship between the effects of a sex-linked gene in males and females. We postulate a simple model of additive gene action in which each gene is responsible for the production of a specified amount of some gene product, irrespective of other genes in the genotype. Consider a sex-linked locus with alleles $B_1, B_2 \ldots$, and write $G^*(B_i)$ for the genotypic value of $B_i$ in males and $G^{**}(B_i B_j)$ for the genotypic value of $B_i B_j$ in females. If the genotypic value is proportional to the amount of gene product, it is natural to suppose that

$$G^{**}(B_i B_j) = G^*(B_i) + G^*(B_j).$$

(6.52)

Since the two genes are statistically independent, it would follow that both the mean and the variance of the character would be twice as large in females as in males.

It would probably be disadvantageous, however, for females to have twice as high a mean value as males for the products of sex-linked loci, and many species have evolved a mechanism of dosage compensation which compensates for the double dose of X chromosomes in females. In most mammals only one X chromosome in each female cell is active, though it is a matter of chance whether it is the paternal or the maternal chromosome; thus female mammals are a mosaic, some cells having an active paternal and others an active maternal X chromosome. With dosage compensation we would therefore predict that

$$G^{**}(B_i B_j) = \tfrac{1}{2} G^*(B_i) + \tfrac{1}{2} G^*(B_j). \tag{6.53}$$

Thus the mean value will be the same in both sexes, but the variance will be twice as large in males as in females. In kangaroos the maternal X chromosome is active and the paternal chromosome inactive in all cells, so that we may write

$$G^{**}(B_i B_j) = G^*(B_j). \tag{6.54}$$

Under this model the mean and the variance will be the same in both sexes. The same would hold if inactivation occurred randomly at conception, so that all X chromosomes were either paternal or maternal. The mechanism of dosage compensation is different in *Drosophila*, but it leads to the prediction (6.53), and Frankham (1977a) has shown experimentally that sex-linked genes affecting abdominal bristle number satisfy this relationship. It therefore seems reasonable to adopt eqn (6.53) as a general model for sex-linked loci affecting characters whose genetic variance is predominantly additive.

We turn now to the covariances between relatives. Sex-linked genes are diploid in females. If we allow the possibility of dominance, the genetic variance can be decomposed into an additive and a dominance component as at an autosomal locus: $V_G^{**} = V_A^{**} + V_D^{**}$. The covariance between two female relatives is $RV_A^{**} + P_2 V_D^{**}$ as at an autosomal locus, but it should be remembered that $R(= \tfrac{1}{2}P_1 + P_2)$ and $P_2$ may have different values at a sex-linked locus (see the top third of Table 3.6). Some numerical values are given in Table 6.11.

Sex-linked genes are haploid in males and the genetic variance at a single sex-linked locus, $V_G^*$, cannot be decomposed further. The covariance between a pair of male relatives is clearly $P_1 V_G^*$, where $P_1$, tabulated in the middle of Table 3.6, is the probability that the relatives are identical by descent at this locus. Numerical values are given in Table 6.11. If there is dosage compensation and no dominance in females ($V_D^{**} = 0$) then $V_G^* = 2V_G^{**}$; if there is no dosage compensation then $V_G^* = \tfrac{1}{2} V_G^{**}$.

Turning to relatives of opposite sex, let the genotypes of the male and female be $B_m'$ and $B_p'' B_m''$ respectively. The genotypic values can be decomposed as follows:

$$\begin{aligned} G^*(B_m') &= m^* + X_1^*(B_m') \\ G^{**}(B_p'' B_m'') &= m^{**} + X_1^{**}(B_p'') + X_2^{**}(B_m'') + X_{12}^{**}(B_p'' B_m''). \end{aligned} \tag{6.55}$$

TABLE 6.11

*Genetic covariances between relatives at a sex-linked locus*

| Relationship | | |
|---|---|---|
| Female relatives | $V_A^{**}$ | $V_D^{**}$ |
| Mother–daughter | 0.5 | 0 |
| Sisters | 0.75 | 0.5 |
| Paternal halfsisters | 0.5 | 0 |
| Maternal halfsisters | 0.25 | 0 |
| Male relatives | $V_G^*$ | |
| Father–son | 0 | |
| Brothers | 0.5 | |
| Paternal halfbrothers | 0 | |
| Maternal halfbrothers | 0.5 | |
| Opposite-sexed relatives | $(V_G^* V_G^{**})^{1/2}$ | |
| Mother–son | 0.707 | |
| Father–daughter | 0.707 | |
| Brother–sister | 0.354 | |
| Paternal halfbrother–halfsister | 0 | |
| Maternal halfbrother–halfsister | 0.354 | |

If the relatives share a pair of identical genes we may suppose without loss of generality that $B'_m = B''_p$. It follows as before that

$$\text{Cov}(G^*, G^{**}) = P_1 E[X_1^*(B'_m) X_1^{**}(B'_m)]. \qquad (6.56)$$

It should be noted that $X_1^*$ and $X_1^{**}$ are different functions, and we cannot proceed further without specifying the relationship between them. The model of dosage compensation embodied in eqn (6.53) implies that $X_1^*(.) = 2X_1^{**}(.)$, so that the covariance is $\frac{1}{2}P_1 V_G^*$. The additive model without dosage compensation in eqn (6.52) implies that $X_1^*(.) = X_1^{**}(.)$ so that the covariance is $P_1 V_G^*$. Both these models assume that there is no dominance in females. The covariance under both models can be expressed in a common form as

$$\text{Cov}(G^*, G^{**}) = 2^{-1/2} P_1 (V_G^* V_G^{**})^{1/2}. \qquad (6.57)$$

This result can be extended to any situation in which $X_1^*(.)$ and $X_1^{**}(.)$ are linearly related. Some numerical results are shown in Table 6.11.

## Threshold characters

The inheritance of some discrete, all-or-nothing characters is difficult to explain under a conventional single locus model, but becomes easy to understand if we suppose that there is an underlying, continuous variable with a critical threshold value, which determines whether or not the character is expressed. Characters which behave in this way are called threshold (or quasi-continuous) characters. The present account is based on Bulmer (1970, Chapter 7). For a recent review see Curnow and Smith (1975).

As an example we shall consider the inheritance of congenital malformations in man. Carter (1965) has shown that the family patterns of the four best-documented malformations (harelip, congenital dislocation of the hip, clubfoot, and pyloric stenosis) are similar. There are differences in detail, and complications arise from differences in the degree of severity of the malformation, from differential incidence in the two sexes, and from deficiencies in the data. Instead of analysing the data for a particular malformation we shall therefore consider an idealized 'typical' malformation with the pattern of inheritance shown in Table 6.12.

TABLE 6.12

*Incidence of congenital malformation among relatives of affected individuals*

| Relationship to affected individual | Coefficient of relationship | Incidence | Tetrachoric correlation |
|---|---|---|---|
| Unrelated general population | 0 | 0.001 | |
| Monozygotic co-twin | 1 | 0.50 | 0.92 |
| Dizygotic co-twin, sib, child | $\frac{1}{2}$ | 0.035 | 0.43 |
| Aunt, uncle, nephew, niece | $\frac{1}{4}$ | 0.005 | 0.22 |
| First cousins | $\frac{1}{8}$ | 0.002 | 0.08 |

To explain the incidence of malformations among relatives under the threshold model, we suppose that the presence or absence of a malformation is determined by an underlying continuous variable, $Y$, and that an individual is affected only if $Y$ exceeds some threshold value, $T$. It may be supposed without loss of generality that $Y$ has been transformed so that it is normally distributed with zero mean and unit variance. The value of the threshold can therefore be calculated from the population incidence of the disease, $P$; thus, if $P = 0.001$, as is the case for the 'typical' malformation in Table 6.12, we may calculate that $T = 3.09$, which is the upper 0.1 per cent point of the standard normal distribution. Furthermore, if we denote the values of the underlying variable in a pair of relatives by $Y$ and $Y^*$, and if the incidence of the malformation in relatives of affected individuals is $P^*$, then

$$\text{Prob}[Y^* > T \,|\, Y > T] = P^*$$

whence

$$\text{Prob}[Y^* > Y \text{ and } Y > T] = PP^*.$$

If we assume that $Y$ and $Y^*$ follow a bivariate normal distribution with correlation $r$, then $r$ can be determined by interpolation in tables of the bivariate normal distribution. This type of correlation is known as *tetrachoric correlation*. Appropriate tables will be found in National Bureau of Standards (1959) and Owen (1962).

Tetrachoric correlations calculated in this way are shown in the last column of Table 6.12. They represent the correlations between the phenotypic values of the underlying continuous variable in different types of relatives. Inspection of these correlations shows that the underlying variable behaves in a very simple way, like a continuous character without dominance or epistasis and with a heritability of about 0.9. The success of the threshold model lies in the simplicity of these results, compared with the complexity of any genetic model which might be constructed to account directly for the incidences among different types of relatives of affected individuals.

Curnow and Smith (1975) consider the concept of an abrupt threshold to be biologically implausible, and they develop an alternative formulation of the model in which an individual with phenotypic value $Y$ has probability $S(Y)$ of developing the trait. They postulate that $S(Y)$, the risk function, is a continuous, increasing function of $Y$ rather than an abrupt jump function. However, they show that, if $Y$ is normally distributed and $S(Y)$ has the shape of a cumulative normal distribution function, then this model is mathematically equivalent to the original threshold model. Thus the simpler threshold model can be used without bothering about the implausibility of the abrupt threshold.

The threshold model can be extended to allow for the many complications which arise in practice. Thus the frequency of a trait often differs in the two sexes; separate threshold values can be calculated for males and females to allow for this. Another important problem, particularly in the study of human genetics in which the environment cannot be randomized between individuals, is that relatives may resemble each other due to common family environment as well as their common genes. To illustrate this problem Table 6.13 shows the data on the incidence of tuberculosis among relatives of tuberculous twins. The fact that the incidence is much higher in monozygotic than in dizygotic co-twins of affected individuals strongly suggests that resistance to tuberculosis is genetically controlled, but it is also likely that both resistance and exposure to infection will be influenced by socio-economic factors which are shared by members of the same family. Under the threshold model the threshold can be calculated as 2.3 from the population incidence of 1.1 per cent. The tetrachoric

TABLE 6.13

*The incidence of tuberculosis among relatives of tuberculous twins (Kallmann and Reisner 1943)*

| Relationship to tuberculous twin | Incidence | Tetrachoric correlation | Corrected correlation |
|---|---|---|---|
| Unrelated general population | 0.011 | | |
| Husband or wife | 0.072 | 0.26 | |
| Parent | 0.166 | 0.52 | 0.26 |
| Dizygotic co-twin, full sib | 0.187 | 0.56 | 0.30 |
| Monozygotic co-twin | 0.615 | 0.93 | 0.67 |

correlations can be calculated in the same way as before, and are shown in the second column of figures. The correlation between spouses can be used to estimate the component of the correlation due to environmental factors. By subtracting this correlation an estimate of the genetic correlation is obtained, and is shown in the last column. It can be concluded that the underlying variable is inherited in a simple manner, without dominance or epistasis, with a heritability of about 0.6; as expected there is also an appreciable environmental correlation between members of the same family.

## Problems

1. Clayton, Morris, and Robertson (1957) studied the numbers of bristles on the fourth and fifth abdominal sternites in *Drosophila*. They found the following variances, where $S$ and $D$ are the sum and the difference of the numbers of bristles on the two sternites:

|  | Males | Females |
|---|---|---|
| Var $(S)$ in original population | 9.2 | 12.5 |
| Var $(D)$ in original population | 3.4 | 5.0 |
| Var $(S)$ within inbred lines derived from original population by sib mating | 4.7 | 5.3 |

From three separate investigations on different types of relatives they obtained the following results on the inheritance of $S$:

| | |
|---|---|
| Intra-sire regression of daughter on dam | $0.27 \pm 0.06$ |
| Intra-sire regression of son on dam† | $0.21 \pm 0.06$ |
| Correlation between half sibs | $0.12 \pm 0.03$ |
| Correlation between full sibs | $0.26 \pm 0.04$ |

† This figure should be multiplied by $(12.5/9.2)^{1/2}$ to correct for the sex difference in the phenotypic variance.

They also estimated the genetic correlation between the numbers of bristles on the two sternites from the results of these experiments; they found that it was not significantly different from 1.

What conclusions can be drawn about the genetic and environmental factors affecting abdominal bristle numbers in *Drosophila*?

2. In an investigation on body weight in poultry (Graybill, Martin, and Godfrey 1956) 22 cockerels were each mated to six hens, and the 12-week body weight of eight male offspring was recorded. The following analysis of variance was obtained:

| Source of variation | Degrees of freedom | Mean square |
|---|---|---|
| Between cockerels | 21 | 0.5629 |
| Between hens within cockerels | 110 | 0.2055 |
| Within clutches | 924 | 0.0924 |

Estimate the correlations between full and half sibs with their standard errors and hence find a confidence interval for the heritability based on the half-sib correlation.

Test whether the data are consistent with the hypothesis that body weight is determined by additive gene action without dominance or environmental correlation between full sibs. Estimate the heritability and its standard error under this hypothesis, and hence find a confidence interval.

3. Paired observations $y_{1i}$ and $y_{2i}$ ($i = 1, 2, \ldots, N$) are made on $N$ pairs of twins. Write down the appropriate analysis of variance table analogous to Table 6.3 and show how the correlation between twins can be estimated from it.

4. Shields (1962) investigated 40 pairs of monozygotic twins brought up apart, a similar number brought up together, and a smaller sample of dizygotic twins. One of the traits studied was a measure of neuroticism based on a self-rating questionnaire. The results are shown below. Carry out analyses of variance on the three groups, and consider the following questions. (i) Does the phenotypic variance differ in the three groups? (ii) Is the correlation lower in monozygotic twins brought up apart than in those brought up together? (iii) What can be inferred about the inheritance of this trait? (Male and female twins have been combined since there is no evidence of a sex difference.)

Monozygotic twins brought up apart

| | | | | | | | | | | | |
|---|---|---|---|---|---|---|---|---|---|---|---|
| 3 | 3 | $5\frac{1}{2}$ | $8\frac{1}{2}$ | 10 | 7 | $6\frac{1}{2}$ | 7 | 14 | 19 | $7\frac{1}{2}$ | 10 |
| 15 | $15\frac{1}{2}$ | 11 | 10 | $10\frac{1}{2}$ | 9 | $11\frac{1}{2}$ | 18 | 13 | 13 | 13 | 11 |
| 18 | 12 | 10 | 8 | $9\frac{1}{2}$ | 7 | 7 | 14 | 17 | 18 | $14\frac{1}{2}$ | 9 |
| $17\frac{1}{2}$ | $14\frac{1}{2}$ | $10\frac{1}{2}$ | $12\frac{1}{2}$ | $9\frac{1}{2}$ | $18\frac{1}{2}$ | $9\frac{1}{2}$ | 16 | 14 | 12 | 17 | 18 |
| $4\frac{1}{2}$ | 5 | 16 | 19 | 8 | 10 | 7 | 8 | 5 | 7 | 16 | 14 |
| 11 | 10 | 13 | 14 | 12 | 12 | 12 | $14\frac{1}{2}$ | $6\frac{1}{2}$ | $4\frac{1}{2}$ | 4 | 8 |
| $9\frac{1}{2}$ | 14 | $8\frac{1}{2}$ | 19 | $13\frac{1}{2}$ | 13 | $13\frac{1}{2}$ | 18 | | | | |

Monozygotic twins brought up together

| | | | | | | | | | | | |
|---|---|---|---|---|---|---|---|---|---|---|---|
| $6\frac{1}{2}$ | $7\frac{1}{2}$ | 8 | 8 | 6 | 7 | 9 | 7 | 8 | 4 | 13 | 3 |
| 10 | 10 | 6 | $11\frac{1}{2}$ | 9 | $7\frac{1}{2}$ | 7 | $7\frac{1}{2}$ | $4\frac{1}{2}$ | 8 | $11\frac{1}{2}$ | 15 |
| 8 | 14 | 8 | 6 | 9 | $11\frac{1}{2}$ | 8 | 14 | 12 | $11\frac{1}{2}$ | 2 | 5 |
| $8\frac{1}{2}$ | 12 | 13 | 7 | 13 | 8 | $4\frac{1}{2}$ | $6\frac{1}{2}$ | 14 | 12 | 8 | $10\frac{1}{2}$ |
| 3 | 17 | 10 | 8 | 14 | $9\frac{1}{2}$ | 9 | 9 | $11\frac{1}{2}$ | 15 | 3 | 3 |
| 7 | $10\frac{1}{2}$ | 12 | $10\frac{1}{2}$ | $13\frac{1}{2}$ | 11 | $10\frac{1}{2}$ | 14 | $11\frac{1}{2}$ | 15 | $15\frac{1}{2}$ | 16 |
| $4\frac{1}{2}$ | 9 | $11\frac{1}{2}$ | $16\frac{1}{2}$ | 7 | $8\frac{1}{2}$ | 10 | $11\frac{1}{2}$ | $5\frac{1}{2}$ | 6 | 1 | 3 |
| $10\frac{1}{2}$ | $10\frac{1}{2}$ | | | | | | | | | | |

Dizygotic twins

| | | | | | | | | | | | |
|---|---|---|---|---|---|---|---|---|---|---|---|
| 7 | 8 | 13 | 8 | 11 | 11 | $13\frac{1}{2}$ | $7\frac{1}{2}$ | 18 | 5 | 13 | $7\frac{1}{2}$ |
| 16 | 13 | $12\frac{1}{2}$ | 9 | $5\frac{1}{2}$ | 4 | 9 | 19 | 10 | $5\frac{1}{2}$ | 9 | 10 |
| 13 | 5 | 10 | $9\frac{1}{2}$ | 20 | 14 | $9\frac{1}{2}$ | 9 | 16 | 5 | $15\frac{1}{2}$ | $16\frac{1}{2}$ |
| $14\frac{1}{2}$ | $5\frac{1}{2}$ | 3 | 10 | 9 | 11 | 13 | 10 | $4\frac{1}{2}$ | 2 | $14\frac{1}{2}$ | 14 |
| 14 | 9 | | | | | | | | | | |

5. In some situations repeated measurements of a character can be made on the same individual, for example in different years or on opposite sides of the body. As a simple model suppose that the genotypic value is the same for all measurements on the same animal, and that the environmental deviation consists of two components, $E = P + T$, where $P$ is a permanent component which arises from environmental factors acting early in life and having a permanent effect on the animal, and $T$ is a temporary component which represents short-term environmental effects and which changes randomly from one measurement to the next. The environmental variance can

be partitioned into permanent and temporary components, $V_E = V_P + V_T$; the correlation between repeated measurements is called the repeatability and is equal to $(V_G + V_P)/V_Y$.

As an example, Turner and Young (1969) present an analysis of the body weight of 86 Merino rams born in the same year and measured at ages 1, 2, and 3. The analysis of variance is shown below. Estimate the components of variance $V_X$, $(V_G + V_P)$ and $V_T$ and hence find the repeatability. What does $V_X$ represent?

| Source of variation | Degrees of freedom | Mean square | Expected mean square |
|---|---|---|---|
| Between years | 2 | 17 940 | $86V_X + V_T$ |
| Between rams | 85 | 202 | $3(V_G + V_P) + V_T$ |
| Residual | 170 | 22 | $V_T$ |

6. From Table 6.7 estimate the genetic and environmental components of variance (in percentage terms) under the following assumptions: (a) $V_{AA} > 0$, $V_{AAA} = V_{AAAA} = \cdots = 0$; (b) $V_{AA} = V_{AAA} \cdots = 0$ for lower order interactions, but $V_{AA\ldots} > 0$ for some high order interaction. Find the predicted correlations between half sibs and between cousins under these models.

7. Verify the results of Table 6.10b. Estimate the genetic and environmental correlations with their standard errors assuming that there is no dominance.

8. Suppose that in males 25 per cent of the phenotypic variance is due to sex-linked genes, 25 per cent to autosomal genes (without dominance or epistasis), and 50 per cent to environmental factors (with the same variance in both sexes). Assume complete dosage compensation and no genotype × sex interaction. Find the regressions of son on father, son on mother, daughter on father, daughter on mother, after standardization of the phenotypic value to have the same mean and variance in the two sexes. (These regressions are the same as the corresponding correlations, which are invariant to standardization.)

# 7 Inbreeding depression and heterosis; the diallel cross

*Many quantitative characters show a decrease in mean value under inbreeding. This phenomenon is called inbreeding depression. The converse phenomenon, an increase in mean value when inbred lines are crossed, is called hybrid vigour or heterosis. In this chapter we shall first consider the theory of inbreeding depression when inbreeding occurs in an outbred population. We shall then discuss the analysis of heterosis from diallel cross experiments in which a number of inbred lines are crossed in all possible combinations; this is an extension of the analysis of crosses between two inbred lines discussed in Chapter 5.*

## Inbreeding depression

In this section we shall discuss the effect of inbreeding on a quantitative character in a population which is initially in Hardy–Weinberg and linkage equilibrium under random mating. The theory can be applied either to sporadic cases of inbreeding in an outbred population, for example the children of marriages between cousins in a human population, or to regular systems of inbreeding, such as continued sib mating.

Inbreeding occurs in the offspring of genetically related parents. An individual's inbreeding coefficient, $f$, was defined on page 39 as the probability that two genes at the same locus are identical by descent. Since these genes must be derived from different parents, an individual's inbreeding coefficient can be identified with the coefficient of consanguinity between his parents, defined as the probability that two homologous genes drawn at random, one from each of the parents, will be identical by descent; if the parents are not themselves inbred their coefficient of consanguinity is half the coefficient of relationship between them, defined on page 74 as $R = \frac{1}{2}P_1 + P_2$.

Inbreeding increases the frequency of homozygotes at the expense of heterozygotes without changing gene frequencies. Write $B_{ij}$ for the $j$th allele at the $i$th locus, and let its frequency be $p_{ij}$. Suppose that the base population with reference to which inbreeding is measured is in Hardy–Weinberg and linkage equilibrium, and that mating occurs at random apart from the factors taken into account in calculating the inbreeding coefficient. If the two genes at the $i$th locus are identical by descent, the probability that they are both $B_{ij}$ is $p_{ij}$; otherwise the two genes are statistically independent. Thus the probability of the ordered genotype $B_{ij} B_{ik}$ is

$$\Pr(B_{ij} B_{ik}) = (1-f)p_{ij} p_{ik} + f \delta_{jk} p_{ij}, \tag{7.1}$$

where $\delta_{jk}$ is the *Kronecker delta function* (cf. eqn 4.18). We also assume that there is no linkage, in which case genes at different loci are statistically independent because of independent assortment.

We shall now consider the mean genotypic value of individuals with inbreeding coefficient $f$ under the probability structure (7.1). We decompose the genotypic value with reference to the random mating base population with $f = 0$, obtaining the fundamental decomposition

$$G = \sum_{S \subseteq N} X_S, \tag{7.2}$$

and we first find the Expected value of a typical component $X_S$. Suppose that $a$ loci are represented once (by either a paternal or a maternal gene) and that $d$ loci are represented twice (by both paternal and maternal genes) in $S$. Thus $X_S$ contributes to the interaction between additive effects at $a$ specified loci and the dominance deviations at $d$ other specified loci; the total interaction is the sum of $2^a$ components, $X_T$, over sets $T$ which are similar to $S$. (See p. 89.)

We first show that, if $a > 0$, then

$$E_f(X_S) = E_0(X_S) = 0; \tag{7.3}$$

$E_f$ denotes an Expectation with respect to the probabilities (7.1). Suppose that the $i$th locus is represented in $S$ by the paternal but not by the maternal gene. Taking the Expectation over this position in the genotype, we find that

$$E_f(X_S) = \sum_i X_S(\ldots B_{ij} \ldots) p_{ij} = E_0(X_S) = 0. \tag{7.4}$$

The dots represent the genes at other positions in $S$ which are held fixed, and it is essential to the argument that these genes are at other loci and are thus statistically independent of $B_{ij}$.

We now consider the case that $a = 0$, $d > 0$. The result (7.4) is still valid in this case provided that there is at least one locus in $S$ (which we take to be the $i$th locus) at which the two genes are not identical by descent. Thus $X_S$ has zero Expectation unless there is identity by descent at all loci in $S$, so that

$$E_f(X_S) = f^d E_0(X_S | \text{all loci in } S \text{ are homozygous}). \tag{7.5}$$

We now define

$$\delta_1 = E_0(D | \text{ all loci are homozygous})$$
$$\delta_2 = E_0(DD | \text{ all loci are homozygous}), \tag{7.6}$$

and so on, where $D$ is the total dominance deviation, $DD$ is the dominance $\times$ dominance interaction, and so on. It follows from eqns (7.3) and (7.5) that

$$E_f(G) = m_0 + f \delta_1 + f^2 \delta_2 + f^3 \delta_3 + \cdots \tag{7.7}$$

where $m_0$ is the mean value in the absence of inbreeding. The main conclusions to be drawn from this result are that any change in the mean under inbreeding is due to dominance, and that in the absence of epistatic interactions between dominance deviations the change in the mean will be proportional to the degree of inbreeding.

Fig. 7.1 illustrates the effect of inbreeding on the average yield in maize. There is a marked linear decline of yield under inbreeding, which can be attributed to the effect of dominance at individual loci without epistatic interactions. The data were obtained as follows. The right-hand point ($f = 1$) is the average of a large number of completely inbred lines of maize, obtained by self-fertilization over many generations. These inbred lines can be regarded as a random sample of lines derived from the original outbred population. (Maize is a naturally outbreeding plant, but can be selfed without difficulty.) The left-hand point ($f = 0$) is the average yield in hybrid crosses between the inbred lines. There were ten two-way crosses (A × B), four three-way crosses ((A × B) × C), and ten four-way crosses ((A × B) × (C × D)), where A, B, C,

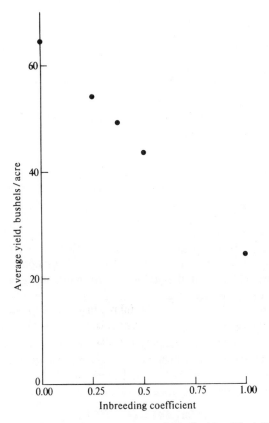

Fig. 7.1. The effect of inbreeding on average yield of maize (Neal (1935), quoted by Crow and Kimura (1970)).

and D are four different inbred lines, but there was no difference in yield between the three types of cross; this is expected since the gametes from different inbred lines represent different gametes from the original outbred population, and the crosses, whether two-way, three-way, or four-way, should, when pooled, reconstruct the original population. Finally, the intermediate points (with $f = \frac{1}{2}, \frac{3}{8}$, and $\frac{1}{4}$) are obtained by random pollination of the two-, three-, and four-way crosses. When a field of two-way hybrids (A × B) is allowed to pollinate at random, the chance that two homologous genes come from the same inbred line is $\frac{1}{2}$, so that $f = \frac{1}{2}$ in the offspring. Likewise, $f = \frac{1}{4}$ in the offspring from a randomly pollinated field of four-way hybrids. For the three-way cross, (A × B) × C, the chance that two homologous genes chosen at random come from C is $\frac{1}{4}$, from A is $\frac{1}{16}$, and from B is $\frac{1}{16}$; thus $f = \frac{3}{8}$ after random pollination.

It is also of interest to consider the effect of inbreeding on the genetic variance. The additive component of the genotypic value is

$$A = \sum_{i=1}^{2n} X_i,$$

and its variance is

$$\mathrm{Var}(A) = \sum_{i=1}^{2n} \mathrm{Var}(X_i) + \sum_{i \neq j} \mathrm{Cov}(X_i, X_j).$$

$\mathrm{Var}(X_i)$ is unaffected by inbreeding since the gene frequencies are unchanged. It is also clear that

$$\mathrm{Cov}(X_i, X_{i+n}) = f\,\mathrm{Var}(X_i)$$
$$\mathrm{Cov}(X_i, X_j) = 0, \quad i \neq j, \bmod n.$$

Hence

$$\mathrm{Var}(A) = (1 + f)V_A, \tag{7.8}$$

where $V_A$ is the additive genetic variance in the absence of inbreeding. Thus inbreeding increases the additive genetic variance by a fraction $f$.

In interpreting this result it should be remembered that it gives the variance among a group of genetically unrelated individuals with inbreeding coefficient $f$. If a large number of inbred or partially inbred lines has been derived from the same base population, then it gives the variance when one individual is chosen at random from each line; equivalently it gives the total variance when all the lines are mixed together and regarded as a single population. This variance can be decomposed into two parts, within and between lines.

It is quite easy to show in the same way that

$$\mathrm{Var}(AA) = (1 + f)^2\, V_{AA}$$
$$\mathrm{Var}(AAA) = (1 + f)^3\, V_{AAA}, \tag{7.9}$$

and so on, and that different components of $G$, such as $A$ and $D$ or $A$ and $AA$, remain uncorrelated under inbreeding. The effect of inbreeding on the dominance deviation and on interactions involving dominance, is more complicated; we shall not pursue this problem further.

Fig. 7.1 represents a rather extreme example of the effect of inbreeding, though many characters in both plants and animals show a similar but smaller response. Note that the mean yield declines under inbreeding (inbreeding depression); conversely, it increases when inbred lines are crossed (hybrid vigour or heterosis). Thus inbreeding depression and hybrid vigour are inverse aspects of the same phenomenon. The term inbreeding depression arises from the fact that the mean value of a character usually declines under inbreeding. Sometimes the opposite happens; for example, Table 7.2 shows that flowering time in tobacco is $4\frac{1}{2}$ days later in inbred varieties than in their cross-bred offspring. In such cases it is convenient to suppose that the character has been redefined so that there is a decline under inbreeding, for example by defining flowering time in days before 31 August rather than in days after 1 July.

When the average value declines linearly, as in Fig. 7.1, it can be concluded that inbreeding depression is due to dominance at individual loci. Two hypotheses have been proposed about the degree of dominance responsible for inbreeding depression. Under the overdominance hypothesis, there exist loci at which the heterozygote is superior to either homozygote; under the dominance hypothesis the heterozygote is at most equal to the better homozygote (complete dominance) and may be inferior to it (partial dominance), but has a tendency to be nearer the better than the worse homozygote (directional dominance). It is important to the plant breeder to be able to distinguish between these hypotheses. If there is only partial or complete dominance it should be possible to combine the best alleles from each locus in a pure-breeding homozygous variety superior to any other genotype; if there is overdominance the best genotype must be heterozygous at all overdominant loci. In the latter case breeding strategy should be aimed at the production of $F_1$ hybrids. Experiments to estimate the average level of dominance using the North Carolina 3 design were discussed at the end of Chapter 5; the evidence is in favour of directional dominance rather than overdominance as the cause of inbreeding depression.

## *The diallel cross

In Chapter 5 we considered the analysis of a cross between two pure lines, but the biologist is often interested in a larger set of lines. A convenient way to study simultaneously the genetic properties of several lines is the mating system known as the diallel cross in which a set of $p$ inbred lines is chosen and crossed in all possible ways. The results may be set out in a square $p \times p$ matrix in which $y_{ij}$ represents the mean value of the offspring of the cross of line $i$ and line $j$. Inbreeding depression and heterosis can be investigated by comparing the hybrid means, $y_{ij}$ with $i \neq j$, with the pure lines, $y_{ii}$. However, we shall begin by describing the statistical analysis of an incomplete diallel table, in which the pure lines are excluded from the analysis.

### The incomplete diallel cross

In the absence of maternal or sex-linked effects, reciprocal crosses ($y_{ij}$ and $y_{ji}$, where the male parent is given first) are equivalent and can be grouped together. Furthermore, in plant breeding interest often lies in the hybrid means rather than in the pure lines, which are excluded from the analysis. In this simple situation we have to analyse the $\frac{1}{2}p(p-1)$ means $y_{ij}$ with $i < j$ as set out below:

$$
\begin{array}{c c c c c c c}
 & 1 & 2 & 3 & \cdots & & p \\
1 & - & y_{12} & y_{13} & \cdots & & y_{1p} \\
2 & & - & - & y_{23} & \cdots & y_{2p} \\
\vdots & & & & & & \\
(p-1) & - & - & - & & - & y_{p-1,p}
\end{array}
\qquad (7.10)
$$

It is assumed that each mean is based on $r$ replications and has sampling variance $V_e$, which can be estimated from the mean square between replicate observations on the same cross.

The obvious statistical model for this design is

$$
E(y_{ij}) = m + g_i + g_j + s_{ij}, \qquad (7.11)
$$

where $m$ is the overall mean, $g_i$ is the main effect of the $i$th line on its hybrid offspring, and $s_{ij}$ is the interaction between the two lines. Sprague and Tatum (1942), who considered crosses between different lines of maize, used the terms *general combining ability* (gca) and *specific combining ability* (sca) for $g_i$ and $s_{ij}$. They defined these terms as follows: 'The term "general combining ability" is used to designate the average performance of a line in hybrid combination.... The term "specific combining ability" is used to designate those cases in which certain combinations do relatively better or worse than would be expected on the basis of the average performance of the lines involved. Among animal breeders the term "nicking" has been used in the same general sense.' It may be added that the general combining ability of plant breeders is similar to the breeding value of animal breeders.

We define $y_{i.}$ as the average value among all crosses in which line $i$ is represented, that is to say the average of the $(p-1)$ entries in either the $i$th row or the $i$th column of the above table of $y_{ij}$s, all of which lie above the diagonal; $y_{..}$ is the overall mean of the $\frac{1}{2}p(p-1)$ entries. The least squares estimators of the parameters (with the constraints that $\sum g_i = 0$ and that $s_{i.} = 0$ for all $i$, where $s_{i.}$ is defined like $y_{i.}$) are

$$\hat{m} = y_{..}$$

$$\hat{g}_i = \frac{(p-1)}{(p-2)}(y_{i.} - y_{..}) \tag{7.12}$$

$$\hat{s}_{ij} = y_{ij} - \hat{m} - \hat{g}_i - \hat{g}_j.$$

The analysis of variance based on the means $y_{ij}$ as units of observation is

| Source of variation | Sum of squares | Degrees of freedom |
|---|---|---|
| General combining abilities | $\dfrac{(p-1)^2}{(p-2)} \sum_i (y_{i.} - y_{..})^2$ | $(p-1)$ |
| Specific combining abilities | By subtraction | $\tfrac{1}{2} p(p-3)$ |
| Total | $\sum_{i<j} (y_{ij} - y_{..})^2$ | |

TABLE 7.1

*Diallel cross on nine inbred lines of maize (Griffing 1956)*

|   | 1 | 2 | 3 | 4 | 5 | 6 | 7 | 8 | 9 |
|---|---|---|---|---|---|---|---|---|---|
| 1 | — | 240 | 260 | 230 | 257 | 242 | 267 | 240 | 300 |
| 2 | — | — | 209 | 217 | 233 | 230 | 267 | 216 | 214 |
| 3 | — | — | — | 184 | 254 | 250 | 269 | 222 | 252 |
| 4 | — | — | — | — | 234 | 214 | 256 | 197 | 281 |
| 5 | — | — | — | — | — | 207 | 272 | 243 | 261 |
| 6 | — | — | — | — | — | — | 262 | 270 | 284 |
| 7 | — | — | — | — | — | — | — | 273 | 302 |
| 8 | — | — | — | — | — | — | — | — | 260 |

Analysis of variance on the above means

| Source of variation | Sum of squares | Degrees of freedom | Mean square |
|---|---|---|---|
| General combining abilities | 18 614 | 8 | 2327 |
| Specific combining abilities | 9126 | 27 | 338 |
| Residual error | | 175 | 151 |

NB: The above analysis relates to hybrid means, each based on 78 observations. The analysis of variance on single observations would have the following components: between crosses (35 d.f.), between blocks (5 d.f.), crosses × blocks (175 d.f.), within crosses and blocks (648 d.f.). The residual error mean square of 151 is the crosses × blocks mean square from this analysis, divided by 78 to convert it to a per mean basis. This is the appropriate error if blocks are treated as random effects with the possibility of genotype × block interaction.

As an example Table 7.1 shows the results of crossing nine lines of maize and the appropriate analysis of variance. The experiment was laid out in a randomized blocks design with six blocks, each cross being replicated 13 times in each block; thus each mean is based on 78 observations. The mean squares for both general and specific combining abilities are highly significant.

We now wish to estimate the relative importance of general and specific combining abilities in determining the result of a cross. We must first decide whether the $p$ lines chosen to form the diallel cross can be thought of as a random sample from a large population of lines (random effects model) or whether they have been chosen deliberately for their own sake (fixed effects model). In the first case we are interested in making inferences about the underlying population of lines, in the latter case interest is restricted to the actual lines in the cross. The Expected mean squares under the two models are shown below:

|  | Random effects model | Fixed effects model |
|---|---|---|
| General combining ability | $(p-2)V_g + V_s + V_e$ | $(p-2)V_g + V_e$ |
| Specific combining ability | $V_s + V_e$ | $V_s + V_e$ |
| Residual error | $V_e$ | $V_e$ |

$V_g$ is defined as $E(g^2)$ under the random effects model and as $\sum g_i^2/(p-1)$ under the fixed effects model, with similar definitions for $V_s$; $V_e$ is the error of replication on a per mean basis. The derivation of these results is left as an exercise for the reader (Problem 3). The variance components can be estimated by equating observed and Expected values. Thus from Table 7.1 we find that

$$\hat{V}_g = 284 \text{ (random effects) or } 311 \text{ (fixed effects)},$$

$$\hat{V}_s = 187 \text{ under both models,}$$

$$\hat{V}_e = 151 \text{ under both models.}$$

Thus specific is less important than general combining ability, but is not negligible. This means that a good, but not a perfect, prediction could be made of a previously untried cross from the behaviour of the two lines involved in the cross.

So far we have analysed the diallel cross by a purely statistical model, and we must now consider its genetic interpretation. This is most easily done under the random effects model. Suppose that the $p$ lines can be regarded as a random sample from a population of lines, and suppose furthermore that this population of lines has been obtained from a random mating population by imposing a system of inbreeding without any selection. Then the population of lines will be in linkage, though not in Hardy–Weinberg, equilibrium, and a single generation of random mating would restore them to both Hardy–Weinberg and linkage equilibrium. Thus the Expectations of crosses with different parents, such as $y_{12}, y_{34}, y_{56}$ and so on, will reproduce the distribution of genotypic values in the original random mating population; hence

$$\text{Var } E(y_{ij}) = 2V_g + V_s = V_G, \qquad (7.13)$$

where $V_G$ is the total genetic variance in the random mating population. Furthermore, two crosses with one parent in common, such as $y_{12}$ and $y_{13}$, will have inherited identical gametes from one parent and independent gametes from the other; they will thus have exactly one pair of identical genes at each locus, like parent and child in a random mating population, so that

$$\text{Cov}(y_{ij}, y_{ik}) = V_g = \tfrac{1}{2}V_A + \tfrac{1}{4}V_{AA} + \tfrac{1}{8}V_{AAA} + \cdots \qquad (7.14)$$

where $V_A$, $V_{AA}$ and so on are the additive, additive $\times$ additive variances and so on in the random mating population under the random mating decomposition. Hence

$$V_s = V_G - (V_A + \tfrac{1}{2}V_{AA} + \tfrac{1}{4}V_{AAA} + \cdots). \qquad (7.15)$$

If there is no epistasis then we can estimate $V_A$ and $V_D$ by $2V_g$ and $V_s$ respectively, but further evidence about the existence of epistasis is desirable. We shall return later to the genetic interpretation of the diallel cross under the fixed effects model.

*Reciprocal effects*

It has been assumed in the above account that reciprocal crosses are equivalent, but this may not be the case if there are maternal or sex-linked effects. We will therefore now consider the analysis of a diallel table in which all entries except diagonal ones are present. The table will have the form:

$$
\begin{array}{ccccc}
— & y_{12} & y_{13} & \cdots & y_{1p} \\
y_{21} & — & y_{23} & \cdots & y_{2p} \\
\vdots & & & & \\
y_{p1} & y_{p2} & y_{p3} & \cdots &
\end{array}
\qquad (7.16)
$$

A natural statistical model for this design is

$$E(y_{ij}) = m + g_i + g_j + s_{ij} + r_{ij}, \qquad (7.17)$$

where $m$, $g_i$, and $s_{ij}$ have the same meaning as before and $r_{ij}$ is the reciprocal effect allowing for the difference between $y_{ij}$ and $y_{ji}$. To accord with this interpretation we postulate that

$$
\begin{aligned}
s_{ji} &= s_{ij} \\
r_{ji} &= -r_{ij}.
\end{aligned}
\qquad (7.18)
$$

To estimate the parameters we impose the constraints

$$\sum_i g_i = 0$$

$$\sum_{j \neq i} s_{ij} = 0 \quad \text{for all } i, \tag{7.19}$$

and we define

$$y_{i.} = \sum_{j \neq i} y_{ij}/(p-1)$$

$$y_{.j} = \sum_{i \neq j} y_{ij}/(p-1) \tag{7.20}$$

$$y_{..} = \sum_{\substack{i,j \\ i \neq j}} y_{ij}/p(p-1).$$

The least squares estimators are

$$\hat{m} = y_{..}$$

$$g_i = \frac{(p-1)}{(p-2)} [\tfrac{1}{2}|y_{i.} + y_{.i}) - y_{..}] \tag{7.21}$$

$$\hat{s}_{ij} = \tfrac{1}{2}(y_{ij} + y_{ji}) - \hat{m} - \hat{g}_i - \hat{g}_j$$

$$\hat{r}_{ij} = \tfrac{1}{2}(y_{ij} - y_{ji}).$$

The analysis of variance is

| Source of variation | Sum of squares | Degrees of freedom |
|---|---|---|
| General combining abilities | $\dfrac{2(p-1)^2}{(p-2)} \sum_i [\tfrac{1}{2}(y_{i.} + y_{.i}) - y_{..}]^2$ | $(p-1)$ |
| Specific combining abilities | By subtraction | $\tfrac{1}{2}p(p-3)$ |
| Reciprocal effects | $\dfrac{1}{2} \sum_{i<j} (y_{ij} - y_{ji})^2$ | $\tfrac{1}{2}p(p-1)$ |
| Total | $\sum_{i \neq j} (y_{ij} - y_{..})^2$ | |

Components of variance can be estimated in the same way as before.

The problem with this approach is that the reciprocal effects, $r_{ij}$, in the statistical model do not have an obvious biological interpretation. It seems better to base the analysis on a model which represents more explicitly the biological reasons for the reciprocal effects. The most likely reason is the presence of maternal effects; they can be represented by the model

$$E(y_{ij}) = m + g_i + g_j + m_j + s_{ij}, \tag{7.22}$$

where $m_j$ is the effect of the $j$th line used as a female parent and the other parameters have the same meaning as before. We postulate that $s_{ji} = s_{ij}$, and we impose the constraints (7.19) together with

$$\sum_j m_j = 0. \tag{7.23}$$

The least squares estimators are

$$\hat{m} = y_{..}$$

$$\hat{g}_i = \frac{(p-1)}{p(p-2)} \left[ (p-1)y_{i.} + y_{.i} - py_{..} \right]$$

$$\tag{7.24}$$

$$\hat{m}_j = \frac{(p-1)}{p} (y_{.j} - y_{j.})$$

$$\hat{s}_{ij} = \tfrac{1}{2}(y_{ij} + y_{ji}) - \hat{m} - (\hat{g}_i + \hat{g}_j) - \tfrac{1}{2}(\hat{m}_i + \hat{m}_j).$$

Complications arise in pushing the analysis further since the design is not orthogonal under this model. It is therefore not possible to write down a simple analysis of variance with additive sums of squares, and several rival methods of estimating the variance components have been proposed (Searle 1971). The simplest method is to equate the observed and Expected values of $\sum \hat{g}_i^2$, $\sum \hat{m}_i^2$, and $\sum \hat{s}_{ij}^2$; this method seems likely to be reasonably efficient.

There are $\tfrac{1}{2}(p-1)(p-2)$ degrees of freedom left to test the adequacy of the model (7.22). Failure of the model to fit might be attributed to interactions between genetic and maternal effects, which can be allowed for by dropping the restriction that $s_{ij} = s_{ji}$ in (7.22); the estimates of $m$, $g_i$, and $m_j$ are unchanged, and $s_{ij}$ is estimated by

$$\hat{s}_{ij} = \hat{y}_{ij} - (\hat{m} + \hat{g}_i + \hat{g}_j + \hat{m}_j). \tag{7.25}$$

The other complication which may have to be considered is sex-linkage. In this case the sex of the offspring as well as that of the parents must be considered. If the male is the heterogametic sex, the model (7.22) is adequate to represent autosomal, sex-linked and maternal effects; in male offspring the sex-linked effects will be confounded with the maternal effects $m_j$, since males inherit their X chromosome from their mother, while in female offspring sex-linked effects will be confounded with autosomal genetic effects $g_i$. If the female is the heterogametic sex, for example in chickens, the model (7.22) will be adequate for male offspring (with sex-linked effects confounded with autosomal effects), but for female offspring we must include an extra term $l_i$ for sex-linked effects which behaves like a paternal effect, since females inherit their X chromosome from their father in this case. It will also be necessary to consider male and female offspring separately if sex × genotype interaction is suspected. See Eisen, Bohren, and McKean (1966) for further discussion; a recent review is given by Hinkelmann (1977).

### The complete diallel cross

So far we have considered an incomplete diallel table in which the values of the parental lines are either missing or ignored. When these values are available they provide valuable evidence about inbreeding depression and heterosis which cannot be inferred from the incomplete table. As an example we shall discuss the data reported by Mather and Jinks (1977) on flowering time (in days after 1 July) of nine inbred lines of tobacco and their $F_1$ hybrids. The experiment was laid out in two blocks, each of which consisted of 81 plots to each of which the 81 progenies of the 9 × 9 matings were assigned at random. Each plot comprised five plants and the datum from each plot is the mean flowering time of the five plants it contained. The flowering time of each of the 81 progenies, averaged over the two blocks, is shown in Table 7.2. The mating types × blocks interaction mean square was 3.858 with 80 degrees of freedom. Since each observation in the table is the average of two plots, the error variance of these observations is estimated as 1.929. (See the note at the bottom of Table 7.1.)

TABLE 7.2

*Flowering time in a 9 × 9 diallel cross in tobacco*

|   | 1 | 2 | 3 | 4 | 5 | 6 | 7 | 8 | 9 |
|---|------|------|------|------|------|------|------|------|------|
| 1 | 38.9 | 26.7 | 39.8 | 34.8 | 25.1 | 29.8 | 35.7 | 33.8 | 25.3 |
| 2 | 23.9 | 27.0 | 25.0 | 23.1 | 21.5 | 26.2 | 23.4 | 20.6 | 20.2 |
| 3 | 34.4 | 26.6 | 48.8 | 29.6 | 25.0 | 31.5 | 36.1 | 24.4 | 26.0 |
| 4 | 36.1 | 23.5 | 31.2 | 34.1 | 23.4 | 29.4 | 27.2 | 22.3 | 25.0 |
| 5 | 26.5 | 23.2 | 26.0 | 25.5 | 26.6 | 27.5 | 27.2 | 20.2 | 24.2 |
| 6 | 28.4 | 24.1 | 30.3 | 31.9 | 24.2 | 27.0 | 27.7 | 22.4 | 24.8 |
| 7 | 36.9 | 24.7 | 41.8 | 33.9 | 30.1 | 29.8 | 37.0 | 24.4 | 29.1 |
| 8 | 26.8 | 19.3 | 27.8 | 22.1 | 19.2 | 18.8 | 22.7 | 15.3 | 21.8 |
| 9 | 25.3 | 23.3 | 24.9 | 24.0 | 22.5 | 21.3 | 27.4 | 19.0 | 25.4 |

The first step in the analysis is to test for reciprocal effects. The mean square for reciprocal effects is

$$\tfrac{1}{2} \sum_{i<j} (y_{ij} - y_{ji})^2 / 36 = 4.019,$$

which gives an $F$ ratio of 2.08 with 36 and 80 degrees of freedom ($P \simeq 0.01$). This result is clearly significant, but it is thought by Mather and Jinks (1977) to arise from bias in the experimental design rather than from biological causes. They write: 'We cannot regard [this result] as clearly demonstrating an extra-nuclear element in the determination of flowering time: true it could reflect such a determinant, but it could arise in other ways too . . .. If the seed for each family was sown in a single seed pan, members of a family, including those plants grown in separate blocks as well as those in the same block, could resemble one another more than they resembled the plants from the reciprocal crosses started off in a different seed pan. This would produce the result observed and later experiments in fact pointed to it as the most likely cause.' We shall therefore assume that reciprocal effects are absent, and we shall take 4.019 as an estimate of the error variance of the observations in Table 7.2.

We shall now analyse the data of Table 7.2 on the assumption that there is no difference between reciprocal crosses. A well-known method of analysing a complete diallel cross has been developed by Hayman and Mather and Jinks (Hayman 1954; Mather and Jinks 1971, 1977). This method is unfortunately based on rather restrictive assumptions which limit its utility. We shall here use the model suggested by Eberhart and Gardner (1966) which is free from these assumptions.

We first define the reference population whose genetic parameters are to be estimated. We start with a fixed set of $p$ homozygous inbred lines. If there are $n$ segregating loci which affect the character, we suppose that the $j$th line is homozygous for the allele $B_{ij}$ at the $i$th locus ($i = 1, \ldots, n; j = 1, \ldots, p$). The alleles $B_{i1}, B_{i2}, \ldots, B_{ip}$ may all be different, but we allow the possibility that some of them may be the same, though it is still convenient to give them different numbers. The reference population is defined to have each of the $p^n$ possible homozygotes present with the same frequency of $p^{-n}$, with a vanishingly small probability of heterozygotes; if we wish to consider a subpopulation conditional on one or more loci being heterozygous, we suppose that at the remaining loci all possible homozygous types occur with equal frequency. We may imagine that this state is reached by crossing the lines in all ways, followed by random mating until linkage equilibrium has been attained, followed by inbreeding (for example, selfing). The appropriate decomposition of the genotypic value is the inbred decomposition defined in eqn (4.24).

We now define the constants

$$a_j = \sum_i A_i(B_{ij} B_{ij})$$

$$d_{jk} = \sum_i D_i (B_{ij} B_{ik}). \tag{7.26}$$

(When $p = 2$, $a_1 = a$ and $d_{12} = d$, where $a$ and $d$ are defined in eqn (5.4).) We assume for simplicity that there is no epistasis. The Expected values of the $j$th pure line and of the cross between lines $j$ and $k$ are

$$E(y_{jj}) = m + a_j$$

$$E(y_{jk}) = m + \tfrac{1}{2}a_j + \tfrac{1}{2}a_k + d_{jk}. \tag{7.27}$$

Before trying to fit this model, we observe that the $a_j$s sum to zero but that there is no comparable constraint on the $d$s. To put the model into a form with the usual constraints, we decompose the $d$s into components

$$d_{jk} = \bar{d} + d_j + d_k + s_{jk} \tag{7.28}$$

where

$$\bar{d} = \sum_{j,k} d_{jk}/p(p-1)$$

$$d_j = \tfrac{1}{2}\left[\sum_k d_{jk}/(p-1) - \bar{d}\right] \tag{7.29}$$

$$s_{jk} = d_{jk} - \bar{d} - d_j - d_k.$$

(Note that $d_{jj} = 0$.) The $d_j$s sum to zero and the $s_{jk}$s sum to zero over either subscript. The quantities $\bar{d}$, $d_j$, and $s_{jk}$ are described by Eberhart and Gardner (1966) as average heterosis, variety heterosis and specific heterosis, respectively.

We can now re-write eqn (7.27) as

$$
\begin{aligned}
E(y_{jj}) &= m + a_j \\
E(y_{jk}) &= (m + \bar{d}) + (\tfrac{1}{2}a_j + d_i) + (\tfrac{1}{2}a_k + d_k) + s_{jk}.
\end{aligned}
\tag{7.30}
$$

Comparing the second equation with eqn (7.11), we can identify $(m + \bar{d})$ in eqn (7.30) with $m$ in eqn (7.11) and $(\tfrac{1}{2}a_j + d_j)$ in eqn (7.30) with $g_j$ in eqn (7.11). In a diallel cross without parental information, like eqn (7.10) and Table 7.1, $\bar{d}$ is confounded with $m$, and $d_j$ is confounded with $a_j$. This provides the genetic interpretation of the incomplete diallel cross when the $p$ lines are regarded as fixed effects, under the inbred decomposition of the genotypic value appropriate to this situation.

We now return to the analysis of the complete diallel cross in Table 7.2 under the model (7.30). The parameters $g_j$ and $s_{jk}$ can be estimated by eqn (7.21) from the incomplete diallel table with the diagonal elements excluded. The parameters in the full model can now be estimated as

$$
\begin{aligned}
\hat{m} &= \bar{y}_{\mathrm{P}} \\
\hat{\bar{d}} &= \bar{y}_{\mathrm{C}} - \bar{y}_{\mathrm{P}} \\
\hat{a}_j &= y_{jj} - \bar{y}_{\mathrm{P}} \\
\hat{d}_j &= \hat{g}_j - \tfrac{1}{2}(y_{jj} - \bar{y}_{\mathrm{P}}),
\end{aligned}
\tag{7.31}
$$

where $\bar{y}_{\mathrm{P}}$ and $\bar{y}_{\mathrm{C}}$ are the mean values of the parental lines and of the crosses respectively. The analysis of variance is shown in Table 7.3. A slight problem arises because the estimates of $a_j$ and $d_j$ are not orthogonal. The sums of squares for these parameters have been calculated by multiplying the sums of squares of the estimates by appropriate factors to make the Expected value of

TABLE 7.3

*Analysis of variance of data in Table 7.2*

| Source of variation | Sum of squares | Degrees of freedom | Mean square |
|---|---|---|---|
| Between varieties $(a_j)$ | $\sum (y_{jj} - \bar{y}_{\mathrm{p}})^2$ | $(p-1) = 8$ | 94.2 |
| Average heterosis $(\bar{d})$ | $(p-1)(\bar{y}_{\mathrm{C}} - \bar{y}_{\mathrm{p}})^2$ | 1 | 162.0 |
| Variety heterosis $(d_j)$ | $\dfrac{4(p-2)}{p} \sum \hat{d}_j^2$ | $(p-1) = 8$ | 18.1 |
| Specific heterosis $(s_{jk})$ | † | $\frac{1}{2}p(p-3) = 27$ | 10.7 |
| Duplicates | $\frac{1}{2} \sum (y_{jk} - y_{kj})^2$ | $\frac{1}{2}p(p-1) = 36$ | 4.0 |

† The sum of squares for specific combining abilities from table like eqn (7.16)

the mean square equal to the residual error variance under the null hypothesis. The sum of squares for specific heterosis has been calculated by finding the sum of squares for specific combining abilities from the table like eqn (7.16) with the diagonal elements excluded.

All the components are significant, but the most important factors are the additive effects of varieties $(a_j)$ and the average heterosis $(\bar{d})$; the average difference in flowering times between inbred varieties and their crossbred offspring is

$$\hat{\bar{d}} = \bar{y}_P - \bar{y}_C = 4.5 \text{ days.}$$

It is also of interest to consider the relationship between $a_j$ and $d_j$; the estimates of these parameters in the nine varieties are shown below:

$\hat{a}_j$  7.8  −4.1  17.7  3.0  −4.5  −4.1  5.9  −15.8  −5.7

$\hat{d}_j$  0.6  −1.6  −5.0  −0.3  −0.2  2.2  0.8  3.6  −0.2

To calculate the correlation between $a_j$ and $d_j$ we first find the following estimates, corrected for sampling bias:

$$\text{Est}(\sum a_j d_j) = \sum \hat{a}_j \hat{d}_j + \tfrac{1}{2} \times 8 \times \hat{V}_e = -78.8$$

$$\text{Est}(\sum a_j^2) = \sum \hat{a}_j^2 - 8\hat{V}_e = 721.9$$

$$\text{Est}(\sum d_j^2) = \sum \hat{d}_j^2 - \frac{8 \times 9}{28} \hat{V}_e = 36.2,$$

where $\hat{V}_e = 4.0$, the residual error variance. The estimated correlation coefficient is −0.49. The negative correlation can be interpreted as meaning that differences among inbred lines $(a_j)$ are magnified compared with the expectation from the behaviour of these lines in hybrid crosses $(2g_j = a_j + 2d_j)$. This may be another manifestation of the instability of homozygotes compared with heterozygotes.

In this section we have only considered a set of pure lines, $L_j$, and the two-way crosses derived from them $(L_j \times L_k)$. Further information can be obtained by selfing these crosses, or by crossing them again to other pure lines in a three-way cross $((L_j \times L_k) \times L_m)$, or by crossing them among themselves in a four-way cross $((L_j \times L_k) \times (L_m \times L_n))$. The analysis follows the same general principles, but we shall not pursue the details here.

## Problems

1. Litter size in mammals may be affected both by inbreeding of the mother (through depression of the number of eggs released or through changes in the intra-uterine environment) and by inbreeding of the young (through increased prenatal mortality). To disentangle these two effects in mice Falconer (1960) inbred a number of lines by brother–sister mating and then crossed lines at random to obtain data on inbred mothers with non-inbred young. He obtained the following results:

| Mating system | Mean litter size |
|---|---|
| Random mating | 8.2 |
| 3 generations of sib-mating | 6.0 |
| 4 generations of sib-mating | 5.8 |
| 2 generations of sib-mating + crossing between lines | 7.5 |
| 3 generations of sib-mating + crossing between lines | 7.3 |
| 3 generations of sib-mating + 2 generations of crossing | 8.5 |

Find the inbreeding coefficients of mother and young in these six cases, and calculate the joint regression of litter size on the two inbreeding coefficients.

2. Define

$$\delta_{1i} = E_0(D \,|\, i\text{th locus is homozygous})$$

where $D_i$ is the dominance deviation at the $i$th locus. Estimate $\delta_{1i}$ in the two allele case defined in Problem 1, Chapter 4, and hence show how an estimate of $V_D$ in an outbred population together with an estimate of inbreeding depression can be used to estimate the number of loci.

3. Show that eqn (7.12) gives the least squares estimators for the model (7.11), and derive the Expected values of the mean squares in the analysis of variance under the random and fixed effects models.

4. Perform an analysis of variance on the data in Table 7.2 ignoring the diagonal elements, using the model (7.17). Find the Expected values of the mean squares under the fixed effects model. Estimate $V_g$ and $V_s$ under the fixed effects model on the assumption that there are no reciprocal effects. If you average reciprocal crosses and analyse the data in the same way as Table 7.1, how will the resulting analysis of variance differ?

5. Show that eqn (7.24) gives the least squares estimators for the model (7.22). Find the Expected values of the sums of squares $\sum \hat{g}_i^2$, $\sum \hat{m}_i^2$, and $\sum \hat{s}_{ij}^2$ under the fixed effects model.

6. Estimate components of variance for additive effects $(a_j)$, variety heterosis $(d_j)$, and specific heterosis $(s_{jk})$ from Table 7.3 under the fixed effects model.

# 8 Normal distribution theory

*In Chapter 6 we considered the correlations between relatives under random mating, but in many contexts, particularly in selection theory, it is necessary to have more detailed information about the joint distribution of phenotypic values in two or more related individuals. Are the regressions linear? Is the variability about them constant?*

*The theory is greatly simplified if it can be assumed that the joint distribution of genotypic (or phenotypic) values among relatives is multivariate normal. Under this assumption all regressions are linear, and the residual error about any regression line is normally distributed with constant variance. The joint distribution is thus completely determined by the covariances. Normality can be justified by the central limit theorem if it can be assumed that the character is determined by a large number of loci with individually small and additive effects.*

*In this chapter we shall first outline the relationship between correlation and regression theory; we shall consider in detail the conditions under which multivariate normality can be assumed; we shall use normal distribution theory to study the effect of assortative mating on correlations between relatives; and we shall finally consider the effect on regressions between relatives of other departures from the conditions under which multivariate normality holds. Normal distribution theory will be used extensively in subsequent chapters, particularly Chapter 9, to study the effect of selection in an outbred population.*

## Correlation and regression

Consider a pair of random variables, $X$ and $Y$, with joint density function $f(x, y)$. The conditional distribution of $Y$ given $X$ is $f(x, y)/\int f(x, y) \, dy$. This conditional distribution has a mean value $E(Y|X = x)$, a variance $\text{Var}(Y|X = x)$ and higher moments, all of which may depend on the conditional value of $X$. The regression of $Y$ on $X$ is the conditional Expectation of $Y$ given $X$, $E(Y|X = x)$. We also define the residual error about the regression as

$$e = Y - E(Y|X = x).$$

This residual error is a random variable which, conditional on $X$, has zero mean and variance $\text{Var}(Y|X = x)$.

We shall now assume that the regression of $Y$ on $X$ is linear so that we may write

$$Y = \alpha + \beta X + e. \tag{8.1}$$

We shall now see how $\alpha$ and $\beta$ can be evaluated in terms of the means, variances, and covariance of $X$ and $Y$. Taking Expected values in eqn (8.1) we find that

$$\alpha = m_Y - \beta m_X. \tag{8.2}$$

Multiplying both sides of eqn (8.1) by $(X - m_X)$ and taking Expected values we find that

$$\beta = \text{Cov}(X, Y)/V_X. \tag{8.3}$$

We can also evaluate the variance of the residual error as follows:

$$\begin{aligned} \text{Var}(e) = \text{Var}(Y - \alpha - \beta X) &= V_Y - 2\beta \,\text{Cov}(X, Y) + \beta^2 \, V_X \\ &= (1 - \rho^2)V_Y \end{aligned} \tag{8.4}$$

where $\rho$ is the correlation coefficient, $\rho = \text{Cov}(X, Y)/(V_X \, V_Y)^{1/2}$. It should be noted that $\beta = \rho$ in the important case when $V_X = V_Y$. If the residual error variance depends on $X$, then eqn (8.4) is to be interpreted as an average variance over all values of $X$.

These results can be generalized to find the regression of $Y$ on a number of variables $X_1, X_2, \ldots, X_k$. We assume that this regression is linear so that we may write

$$Y = \alpha + \beta_1 \, X_1 + \beta_2 \, X_2 + \cdots + \beta_k \, X_k + e. \tag{8.5}$$

Taking Expected values we find that

$$\alpha = E(y) - \sum \beta_i E(X_i). \tag{8.6}$$

Multiplying both sides of eqn (8.5) by $X_i - E(X_i)$ and taking Expected values for $i = 1, \ldots, k$ we obtain $k$ equations which can be arranged in matrix form to give:

$$\mathbf{V}\boldsymbol{\beta} = \mathbf{c}. \tag{8.7}$$

In this equation $\mathbf{V}$ is the covariance matrix of the $X_i$s, $V_{ij} = \text{Cov}(X_i, X_j)$; $\boldsymbol{\beta}$ is the column vector of the $\beta_i$s; and $\mathbf{c}$ is the column vector of the covariances of $Y$ with the $X_i$s, $c_i = \text{Cov}(Y, X_i)$. Hence

$$\boldsymbol{\beta} = \mathbf{V}^{-1} \mathbf{c}. \tag{8.8}$$

Evaluating the residual error variance in the same way as before we find that

$$\text{Var}(e) = \text{Var}(Y) - \mathbf{c}^T \mathbf{V}^{-1} \mathbf{c}. \tag{8.9}$$

In the general case when the regression is not necessarily linear, the above results can be interpreted as giving the best linear approximation to the true regression. Consider a pair of variables, $X$ and $Y$, and suppose that the regression of $Y$ on $X$, $E(Y \mid X = x)$, is a non-linear function of $x$. Among linear functions, $Y = \alpha + \beta X$, we seek the line which has minimum mean square error, $E(e^2)$, where $e = Y - \alpha - \beta X$. Writing

$$e = (Y - m_Y) - \beta(X - m_X) + (m_Y - \alpha - \beta m_X),$$

squaring, and taking the Expected value we find that

$$E(e^2) = V_Y - 2\beta \operatorname{Cov}(X, Y) + \beta^2 V_X + (m_y - a - \beta m_X)^2.$$

It is easily shown that this function of $a$ and $\beta$ has a unique minimum when these parameters take the values given in eqns (8.2) and (8.3), and that this minimum value is given by eqn (8.4). This type of regression is called the *linear mean square regression*. The argument can be generalized to the multivariate case, and it can be shown that eqns (8.6) and (8.8) define the linear mean square regression, with $E(e^2)$ given by the right hand side of eqn (8.9).

It is important to distinguish between these two types of regression. The distinction is most clearly seen in the distribution of the error term, $e$. If eqn (8.1) represents the true regression of $Y$ on $X$, then the conditional Expectation of $e$ given $X$ is zero for all $X$. If the true regression is non-linear so that eqn (8.1) represents only the best linear approximation to it, then the Expectation of $e$ over $X$ and $Y$ is zero, but the conditional Expectation of $e$ given $X$ is not zero for all $X$ since it represents the difference between the linear mean square regression and the true regression.

### Multivariate normality

Assume random mating and consider a character determined by $n$ loci without epistasis, so that we may write

$$G = m + G_1 + G_2 + \cdots + G_n,$$

where $G$ is the genotypic value and $G_i$ is the contribution to it from the $i$th locus. Under random mating the contributions from different loci will be statistically independent, whether or not the loci are linked, so that by the central limit theorem their sum will become asymptotically normal as $n$ increases.

Now let us decompose the contributions at each locus into additive and dominance components, $G_i = A_i + D_i$, so that we may write

$$G = m + (A_1 + D_1) + (A_2 + D_2) + \cdots + (A_n + D_n)$$

$$= m + A + D.$$

The pairs of random variables $(A_i, D_i)$ at different loci will be statistically independent (whether or not the loci are linked), so that by the bivariate form of the central limit theorem $A$ and $D$ will become bivariate normal when $n$ is large. Since they are uncorrelated they must become independent normal variates as $n$ increases by a well-known property of the bivariate normal distribution.

We now consider two related individuals with genotypic values $G'$ and $G''$ which can be written

$$G' = m + G'_1 + G'_2 + \cdots + G'_n$$

$$G'' = m + G''_1 + G''_2 + \cdots + G''_n.$$

Under random mating and in the absence of linkage the pairs of random variables $(G_i', G_i'')$ at different loci will be independent of each other because unlinked loci assort independently; in consequence $G'$ and $G''$ will become bivariate normal for large $n$.

As before we can decompose the contributions at each locus into additive and dominance components. In the absence of linkage the tetrads $(A_i', A_i'', D_i', D_i'')$ at different loci will be statistically independent. By the multivariate central limit theorem, the additive effects and the dominance deviations in the two individuals, $A'$, $A''$, $D'$, and $D''$, will become multivariate normal for large $n$. Furthermore, the additive effects and the dominance deviations are uncorrelated with each other and are thus asymptotically independent. Thus $A'$ and $A''$ are asymptotically bivariate normal with zero mean, variance $V_A$ and covariance $RV_A$; $D'$ and $D''$ will asymptotically follow an independent bivariate normal distribution with zero mean, variance $V_D$ and covariance $P_2 V_D$.

These results can be extended in the obvious way to any number of related individuals. Under random mating and in the absence of linkage the joint distribution of the genotypic values will in the limit become multivariate normal. The joint distribution of the additive effects will be multivariate normal with zero mean and with covariance matrix $\mathbf{R} V_A$, where $\mathbf{R}$ is the matrix of coefficients of relationship between pairs of relatives. The joint distribution of the dominance deviations will be independently multivariate normal with zero mean and with covariance matrix $\mathbf{P}_2 V_D$, where $\mathbf{P}_2$ is the matrix of $P_2$ values of pairs of relatives.

It was necessary to assume that there was no linkage in showing that the genotypic values of two or more related individuals are asymptotically multivariate normal. Consider for example a pair of related individuals. In the presence of linkage the pairs $(G_i', G_i'')$ at different loci will not in general be independent of each other. The reason is that knowledge of this pair of values at a particular locus provides some additional information about the number of identical genes at that locus, which provides information about the number of identical genes at a second linked locus, which in turn provides information about the joint distribution of the pair of genotypic values at the second locus; thus if loci $i$ and $j$ are linked one would expect the absolute deviations $|G_i' - G_i''|$ and $|G_j' - G_j''|$ to be positively correlated. The exception to this rule occurs when the number of identical genes is known for sure from the relationship, that is to say for unrelated individuals ($P_0 = 1$), for parent and child ($P_1 = 1$), and for identical twins ($P_2 = 1$). In these cases bivariate normality of the genotypic values, and of their additive and dominance components, will hold whether or not there is linkage. (Remember that the population is supposed to be in linkage equilibrium under random mating.) This result can be extended to the joint distribution of father, mother and child, and to any number of identical sibs, but it will not be true when there is linkage if the individuals are related in any other way.

The most important properties of multivariate normality are that all regressions are linear and that the residual error about the regression line is normally distributed with constant variance. Furthermore, the regression line and the residual error variance can be computed explicitly from the means, variances and covariances of the variables as shown in the last section.

The most important regressions are those of an individual on his parents and other direct ancestors. We write $Y$ for phenotypic values, and suppose that $Y = G + E$, where $E$ is distributed normally and independently of $G$ with zero mean and variance $V_E$, so that multivariate normality is preserved. Denote the phenotypic values of a child and of his mother and father as $Y_C$, $Y_M$, $Y_F$. The regression of the child on one parent, say the mother, is

$$Y_C = (1 - \tfrac{1}{2}h^2)m + \tfrac{1}{2}h^2\,Y_M + e$$
$$\mathrm{Var}(e) = (1 - \tfrac{1}{4}h^4)V_Y,$$

(8.10)

where $h^2 = V_A/V_Y$. The regression of the child on both parents is

$$Y_C = (1 - h^2)m + \tfrac{1}{2}h^2\,Y_M + \tfrac{1}{2}h^2\,Y_F + e$$
$$\mathrm{Var}(e) = (1 - \tfrac{1}{2}h^4)V_Y.$$

(8.11)

The regression of the child on the mid-parental value, $X = \tfrac{1}{2}(Y_M + Y_F)$, considered as a single variable, is

$$Y_C = (1 - h^2)m + h^2\,X + e$$
$$\mathrm{Var}(e) = (1 - \tfrac{1}{2}h^4)V_Y,$$

(8.12)

as is obvious from eqn (8.11).

The regression of an individual's phenotypic value on the phenotypic values of his parents, grandparents, and more remote ancestors can be calculated in a similar way, but the results are rather complicated. Much simpler results are obtained by considering separately the regressions of the three components of the phenotypic value, $A$, $D$, and $E$. All three components are distributed independently of each other in the population, and both $D$ and $E$ are distributed independently of the values of $A$, $D$, or $E$ in any direct ancestor or descendant. The regression of the additive component of a child on the additive components of his parents is

$$A_C = \tfrac{1}{2}A_M + \tfrac{1}{2}A_F + e$$
$$\mathrm{Var}(e) = \tfrac{1}{2}V_A.$$

(8.13)

Furthermore, $e$ is distributed independently of any other ancestors. Let $G$ be an ancestor on the maternal side $k$ generations before C. The correlations of $A_G$ with $A_C$, $A_M$, and $A_F$ are $\tfrac{1}{2}^k$, $\tfrac{1}{2}^{k-1}$, and 0. Thus $e = A_C - \tfrac{1}{2}(A_M + A_F)$ is uncorrelated with $A_G$, and is consequently independent of $A_G$ because of multivariate normality. It follows that eqn (8.13) represents the regression of $A_C$ on the additive components of all C's direct ancestors.

The results of the last paragraph, though very simple, are of great importance in studying the effects of selection as we shall see in later chapters. In the next section we shall use these results to study the effects of assortative mating.

## Assortative mating

A problem which sometimes arises in the analysis of data from natural populations is that individuals do not mate at random with respect to some characters but that there is a tendency for like to mate with like. The effect of such assortative mating on the phenotypic variance and on correlations between relatives was first investigated by Fisher (1918) in a notoriously difficult paper. In this section we shall first derive Fisher's results in a simpler way based on the normal distribution theory developed above.

### *Effect of assortative mating on the variance*

We shall consider a form of assortative mating which acts through introducing a correlation between the phenotypic values of mates but leaves their marginal distributions unchanged, so that there is no selection in the strict sense. We assume that in generation $t$ the phenotypic value, $Y$, is normally distributed with mean $m$ and variance $V_Y(t)$ (this assumption will be justified in a moment), and we suppose that individuals mate assortatively in such a way that the phenotypic values of parents, say $Y_F$ and $Y_M$, follow a bivariate normal distribution with mean $m$, variance $V_Y(t)$ and correlation $\rho$. We also suppose that assortative mating starts at time $t = 0$, and that at that time the population is in Hardy–Weinberg and linkage equilibrium, and that the normal distribution theory of the last section holds. Thus in generation 0 we may write

$$Y = m + A + D + E \tag{8.14}$$

where $A$, $D$, and $E$ are independently and normally distributed with zero means and with variances $V_A$, $V_D$, and $V_E$.

We now consider the joint distribution of $A$ and $D$ of an individual in generation $t$ conditional on the values of $A$, $D$, and $E$ of all his direct ancestors back to generation 0. Since assortative mating acts only by introducing a correlation between the phenotypic values of mates, it is clear that the conditional distribution with all the ancestral phenotypic values held fixed is unaffected by assortative mating and is the same as it would have been under random mating. It follows that the dominance deviation, $D$, in any generation remains normally distributed with variance $V_D$, independently of the dominance deviations or additive components of all ancestors. Likewise the regression of the additive component of an individual on those of his parents is

$$A_C = \tfrac{1}{2}A_F + \tfrac{1}{2}A_M + e, \tag{8.15}$$

where the residual error, $e$, is normally distributed with zero mean and variance $\frac{1}{2}V_A$, independently of the additive components or dominance deviations of any of the individual's direct ancestors.

Thus the distributions of $D$ and $E$ are unaffected but the distribution of $A$ will change under assortative mating. We suppose that in generation $t$ $A$ is normally distributed with zero mean and variance $V_A(t)$; it follows from the argument of the previous paragraph that $A$, $D$, and $E$ are independently distributed. Hence the phenotypic value is normal (as already postulated) with mean $m$ and variance

$$V_Y(t) = V_A(t) + V_D + V_E. \tag{8.16}$$

The regression of $A$ on $Y$ is

$$A = h^2(t)(Y - m) + e, \tag{8.17}$$

where $h^2(t) = V_A(t)/V_Y(t)$ is the heritability in generation $t$, and where the residual, $e$, is normal with zero mean and with variance

$$\mathrm{Var}(e) = V_A(t)(1 - h^2(t)). \tag{8.18}$$

If $Y_F$ and $Y_M$ are the phenotypic values of a mating couple, their additive components can be represented by

$$
\begin{aligned}
A_F &= h^2(t)(Y_F - m) + e_F \\
A_M &= h^2(t)(Y_M - m) + e_M,
\end{aligned}
\tag{8.19}
$$

where $e_F$ and $e_M$ are independently and normally distributed with zero mean and with variance given by eqn (8.18). (This argument depends on the assumption that assortative mating acts through phenotypic values and only depends on genotypic values or components of them through their relationship with phenotypic values.) Hence $A_F$ and $A_M$ have a bivariate normal distribution with zero mean, variance $V_A(t)$ and correlation $ph^2(t)$. It now follows from eqn (8.15) that the additive component in the next generation is normally distributed with zero mean and with variance

$$V_A(t + 1) = 0.5V_A(t)(1 + ph^2(t)) + 0.5V_A, \tag{8.20}$$

where $V_A$ is the additive variance under random mating.

We have thus justified the assumption of normality for the additive components under the model of assortative mating postulated here. We have also found a recurrence relationship for the additive genetic variance in successive generations of assortative mating. Denoting steady state values by 'hats', we find that

$$\hat{V}_A(1 - p\hat{h}^2) = V_A. \tag{8.21}$$

To use this equation to find $\hat{V}_A$ given the variances under random mating, write

$\hat{h}^2 = \hat{V}_A/(\hat{V}_A + V_D + V_E)$, and hence find a quadratic equation for $\hat{V}_A$ with the unique positive root

$$\hat{V}_A = \frac{(V_A - V_D - V_E) + [(V_A + V_D + V_E)^2 - 4\rho V_A(V_D + V_E)]^{1/2}}{2(1-\rho)}. \quad (8.22)$$

Table 8.1 shows some numerical examples. It will be seen that an appreciable increase in the additive genetic variance can occur under positive assortative mating. This increase is due to the correlations between loci (departures from linkage equilibrium) which are induced by assortative mating.

TABLE 8.1

*The steady-state additive genetic variance, $\hat{V}_A$, under assortative mating when the additive variance under random mating, $V_A$, is 100*

| $V_D + V_E$ | $\rho$ | | |
|---|---|---|---|
| | 0.25 | 0.5 | 0.75 |
| 0 | 133 | 200 | 400 |
| 100 | 115 | 141 | 200 |
| 500 | 105 | 110 | 117 |

### Correlations between relatives

The correlations between relatives under assortative mating are affected both directly by the increase in the additive genetic variance and indirectly by the correlation between mates. We note the covariances between the additive components and phenotypic values of mates, which can be derived from eqn (8.19):

$$\text{Cov}(Y_F, Y_M) = \rho \hat{V}_Y$$
$$\text{Cov}(Y_F, A_M) = \rho \hat{h}^2 \hat{V}_Y = \rho \hat{V}_A \quad (8.23)$$
$$\text{Cov}(A_F, A_M) = \rho \hat{h}^2 \hat{V}_A.$$

We shall now derive the correlations between parent and child, between grandparent and grandchild, and between sibs.

To find the parent–child correlation, multiply both sides of eqn (8.15) by $Y_F$ and take Expected values; then

$$E(A_C Y_F) = 0.5E(A_F Y_F) + 0.5E(A_M Y_F) = 0.5(1 + \rho)\hat{V}_A. \quad (8.24)$$

The child's dominance deviation is uncorrelated with the parent's phenotypic value, so that this is also the total covariance, $\text{Cov}(Y_C, Y_F)$. It is worth noting, however, that

$$E(Y_C A_F) = E(A_C A_F) = 0.5E(A_F^2) + 0.5E(A_M A_F) = 0.5(1 + \rho \hat{h}^2)\hat{V}_A. \quad (8.25)$$

To find the grandparent–grandchild correlation suppose that G is the father of F. Writing $A_C = \frac{1}{2}(A_F + A_M) + e$, multiplying by $Y_G$ and taking Expected values,

$$E(A_C Y_G) = \tfrac{1}{2}E(A_F Y_G) + \tfrac{1}{2}E(A_M Y_G).$$

$E(A_F Y_G)$ is given by eqn (8.24). $A_M$ is only correlated with $Y_G$ through the marital correlation of $Y_M$ and $Y_F$; writing $A_M = \rho \hat{h}^2 Y_F + e$, we find that

$$E(A_M Y_G) = \rho \hat{h}^2 E(Y_F Y_G).$$

Hence

$$\text{Cov}(Y_C, Y_G) = E(A_C Y_G) = 0.25(1 + \rho)(1 + \rho \hat{h}^2)\hat{V}_A. \qquad (8.26)$$

To find the correlation between sibs suppose that C and C* are two children of F and M. Writing $A_C = \frac{1}{2}(A_F + A_M) + e$, multiplying by $Y_{C*}$ and taking Expected values

$$E(A_C Y_{C*}) = \tfrac{1}{2}E(A_F Y_{C*}) + \tfrac{1}{2}E(A_M Y_{C*})$$
$$= \tfrac{1}{2}(1 + \rho \hat{h}^2)\hat{V}_A, \text{ from eqn (8.25)}.$$

The covariance between the dominance deviations is unaffected by assortative mating. Hence

$$\text{Cov}(Y_C, Y_{C*}) = \tfrac{1}{2}(1 + \rho \hat{h}^2)\hat{V}_A + \tfrac{1}{4}V_D. \qquad (8.27)$$

TABLE 8.2

*Covariances between relatives under assortative mating*

| Relationship | $\hat{V}_A$ | $V_D$ |
|---|---|---|
| Parent–child | $0.5(1 + \rho)$ | 0 |
| Child–mid-parent | $0.5(1 + \rho)$ | 0 |
| Grandparent–grandchild | $0.25(1 + \rho)(1 + \rho \hat{h}^2)$ | 0 |
| Sibs | $0.5(1 + \rho \hat{h}^2)$ | 0.25 |
| Half-sibs | $0.25(1 + \rho \hat{h}^2)^2$ | 0 |
| Uncle–nephew | $0.25(1 + \rho \hat{h}^2)^2$ | $0.125\rho \hat{h}^2$ |
| Cousins | $0.125(1 + \rho \hat{h}^2)^3$ | $0.0625(\rho \hat{h}^2)^2$ |
| Double first cousins† | $0.25(1 + \rho \hat{h}^2)^2$ | 0.0625 |

† This result differs from Fisher's (1918) result.

Correlations for other relatives can be found in a similar way, and are tabulated in Table 8.2; the reader is recommended to draw the appropriate pedigrees before trying to reproduce these results.

To illustrate the use of these results in interpreting observed correlations we shall consider the data of Pearson and Lee (1903) on the inheritance of height in over 1000 human families. The correlations, which were first analysed by Fisher (1918), are shown in Table 8.3. They are clearly unaffected by sex, so

TABLE 8.3

*Correlations for height (Pearson and Lee 1903)*

| Relationship | Correlation coefficient $\pm$ standard error |
|---|---|
| Husband–wife | $0.280 \pm 0.028$ |
| Father–son | $0.514 \pm 0.022$ |
| Father–daughter | $0.510 \pm 0.019$ |
| Mother–son | $0.494 \pm 0.024$ |
| Mother–daughter | $0.507 \pm 0.021$ |
| Average parent–child | $0.506 \pm 0.011$ |
| Brother–brother | $0.511 \pm 0.042$ |
| Brother–sister | $0.553 \pm 0.019$ |
| Sister–sister | $0.537 \pm 0.033$ |
| Weighted average sib–sib | $0.543 \pm 0.015$ |

Note: the standard errors have been calculated from the probable errors given by Pearson and Lee (1903); they are only approximate since the correlations were calculated as product moment correlations from tables of paired observations.

that we need only consider the average parent–child and sib–sib correlations. The predicted parent–child correlation is

$$r_{P-C} = \tfrac{1}{2}(1 + \rho)\hat{h}^2$$

so that the heritability can be estimated as

$$\hat{h}^2 = 2r_{P-C}/(1 + \rho) = 2 \times 0.506/1.28 = 0.791 \pm 0.017.$$

(The husband–wife correlation has been regarded as known in finding the standard error since it is based on the same data as the parent–child correlation.)

It is often more convenient to estimate the heritability from the regression of child on mid-parent rather than from the parent–child correlation. It will be found from Table 8.2 that the regression coefficient on the mid-parent is $\hat{h}^2$, so that the observed regression can be used directly as an estimate of the heritability. The joint regressions of sons and daughters on their fathers and mothers are given by Pearson and Lee (1903). After standardizing the four height distributions to have zero mean and unit variance, the regressions are

son's deviation = 0.407 (father's deviation) + 0.379 (mother's deviation)

daughter's deviation = 0.399 (father's deviation) + 0.395 (mother's deviation).

There is clearly no effect of the sex of either parent or child, and by averaging the above regressions we find that

child's deviation = 0.790 (average parental deviation).

The estimate of the heritability is identical with the value obtained from the parent–child correlations.

The sib correlation can now be used to estimate the proportion of the phenotypic variance attributable to dominance. The predicted sib correlation is

$$r_{\text{sibs}} = \tfrac{1}{2}(1 + \rho\hat{h}^2)\hat{h}^2 + \tfrac{1}{4}V_D/\hat{V}_Y,$$

so that this proportion can be estimated as

$$V_D/\hat{V}_Y = 4r_{\text{sibs}} - 2(1 + \rho\hat{h}^2)\hat{h}^2 = 0.240 \pm 0.077.$$

(The 'delta technique' has been used in finding the standard error, treating $\rho$ as fixed as before.) The total variance can thus be partitioned as follows in percentage terms:

|  |  |
|---|---|
| Additive genetic | 79 per cent |
| Dominance | 24 per cent |
| Environmental (by subtraction) | (−3 per cent) |
| Total | 100 per cent |

On the face of it there would seem to be a negligible contribution to height from environmental factors and quite a large contribution from dominance, but we must bear in mind the possibility that the correlation between sibs is inflated by their similar environments; if this is so then the contribution of dominance has been overestimated and that of the environment underestimated.

The exact results derived above depend on the assumptions underlying the ideal normal theory (an effectively infinite number of unlinked loci without epistasis) together with the assumption that the joint phenotypic distribution of mates is bivariate normal. It seems likely that this theory will remain approximately valid for most situations likely to be of practical interest, but it is difficult to justify this remark rigorously. We shall discuss this question further in the next section.

## *Departures from normal theory

Normal distribution theory developed in the second section of this chapter is an ideal theory which rests on the assumptions of an infinite number of loci without epistasis or linkage and with a normally distributed environmental deviation. We shall now consider the effects of departures from these assumptions on regressions between relatives, and in particular on the linearity of these regressions. It should be remembered that if the true regression is linear, then it can be calculated by the theory of the first section, regardless of normality. When the true regression is non-linear, the linear mean square regression of the first section may provide an acceptable approximation to it, if the departure from linearity is not too large.

*Additive gene action*

Consider a character determined by $n$ loci each with two alleles, $B$ and $b$; suppose that these alleles contribute 1 and 0 respectively to the character value, without dominance, and that their gene frequencies are $p$ and $q(=1-p)$ at each locus. Write $G_i$ for the contribution to the genotypic value from the $i$th locus; thus $G_i = 0$, 1, or 2 according as the unordered genotype at the $i$th locus is $bb$, $Bb$, or $BB$. The genotypic value is $G = \sum G_i$; it follows a binomial distribution $B(p, 2n)$ with mean $2np$ and variance $2npq$. The purpose of studying this model is to gain insight into the behaviour of regressions between relatives under additive gene action (no dominance or epistasis) with a limited number of loci.

Denote the genotypic values in father and child by $G_F$ and $G_C$. Since the covariance between them is $\frac{1}{2}V_G$, the mean square linear regression of $G_C$ on $G_F$ is

$$G_C = np + \tfrac{1}{2}G_F + e$$
$$E(e^2) = 1\tfrac{1}{2}npq. \tag{8.28}$$

We shall now find the true regression.

Denote the contributions from the $i$th locus in father and child by $G_{iF}$ and $G_{iC}$ respectively. It is clear that $G_{iC}$ is independent of $G_{jF}$ for $i \neq j$. Since the child receives one of the two paternal genes at random, and receives $B$ or $b$ from his mother with probabilities $p$ and $q$, it follows that

$$E(G_{iC}|G_{1F}, G_{2F}, \ldots, G_{nF}) = E(G_{iC}|G_{iF}) = p + \tfrac{1}{2}G_{iF}. \tag{8.29}$$

Summing over $i$ we find that

$$E(G_C|G_{1F}, G_{2F}, \ldots, G_{nF}) = np + \tfrac{1}{2}G_F. \tag{8.30}$$

It follows that

$$E(G_C|G_F) = np + \tfrac{1}{2}G_F. \tag{8.31}$$

Thus the mean square linear regression is in fact the true regression.

To find the variance about this regression line we first observe that

$$\text{Var}(G_{iC}|G_{iF}) \begin{array}{ll} = pq & \text{if } G_{iF} = 0 \text{ or } 2 \\ = pq + 0.25 & \text{if } G_{iF} = 1. \end{array} \tag{8.32}$$

The reason is that if the father is homozygous the only variability arises from the maternal contribution, whereas if the father is heterozygous there is additional variability from the paternal contribution. We also note that

$$\text{Cov}(G_{iC}, G_{jC}|G_{iF}, G_{jF}) = 0 \quad \text{if } i \neq j. \tag{8.33}$$

If the loci are unlinked this follows from independent assortment. When the loci are linked a correlation may arise if the numbers of pairs of identical genes at the two loci are not independent. However, this complication does not arise in the case of parent and child since they always have exactly one pair of identical genes at every locus. Summing over loci we find that

$$\text{Var}(G_C | G_{1F}, G_{2F}, \ldots, G_{nF}) = npq + 0.25 H_F \tag{8.34}$$

where $H_F$ is the number of heterozygous loci in the father. We must now evaluate the average number of heterozygous loci conditional on $G_F$. The conditional probability that a particular locus is heterozygous is

$$2G_F(2n - G_F)/2n(2n - 1).$$

(This is the hypergeometric probability of obtaining one red ball and one black ball in sampling two balls without replacement from an urn containing $G_F$ red balls and $2n - G_F$ black balls; red and black balls represent $B$ and $b$ genes.) The Expected value of $H_F$ is $n$ times this quantity. Hence

$$\text{Var}(G_C | G_F) = npq + 0.25 G_F(2n - G_F)/(2n - 1). \tag{8.35}$$

If we average this residual variance over $G_F$ we find that its average value is $1.5npq$, in agreement with eqn (8.28). However, the variance does change slightly with $G_F$ because of changes in the heterozygosity of the father. When $p = q = \frac{1}{2}$, the ratio of the maximum value of eqn (8.35) (when $G_F = n$) to the average value is $(6n - 2)/(6n - 3)$. Thus the effect is likely to be negligible unless the number of loci is rather small.

The above arguments can easily be extended to find the joint regression of the child's genotypic value, $G_C$, on the genotypic values of both father and mother, $G_F$ and $G_M$. Since the contributions from the two parents are independent and additive, we find that

$$E(G_C | G_F, G_M) = \tfrac{1}{2}G_F + \tfrac{1}{2}G_M \tag{8.36}$$

$$\begin{aligned} \text{Var}(G_C | G_F, G_M) = {} & 0.25 G_F(2n - G_F)/(2n - 1) \\ & + 0.25 G_M(2n - G_M)/(2n - 1). \end{aligned} \tag{8.37}$$

Thus the joint regression of the child on both parents is linear, and the variance about the line depends on the heterozygosity of the two parents. These results hold whether or not there is linkage.

We consider next the regression of the child on both parents and on some or all of his grandparents and other more remote direct ancestors. If we know the number of $B$ genes at the $i$th locus in the father, our assessment of the probability that the father will transmit a $B$ gene to his child is unaffected by knowledge about the genotypes of grandparents, greatgrandparents, and so on. Thus $G_{iC}$ depends only on $G_{iF}$ and $G_{iM}$ and is independent of the genotypic values of the grandparents or any other direct ancestors. Hence eqn (8.36) also represents the regression on both parents plus any other ancestors. However, the variance about the regression line and other characteristics of the distribution of the residual error may be affected by knowing the genotypic values of other ancestors. Consider, for example, a character determined by two loci with recombination fraction $r$. If $G_F = G_M = 2$, we find from eqn (8.37) that the residual error variance is $2/3$. If we knew in addition that both grandfathers had $G = 4$ and both grandmothers $G = 0$, then the probabilities that the father's gametes carry 0, 1, or 2 $B$ genes are $(1 - r)/2, r$, and $(1 - r)/2$ respectively, and likewise for the mother's gametes. Thus the mean and the variance of the number of $B$ genes in the gametes of either parent are 1 and $(1 - r)$ respectively. Thus $E(G_C) = 2$, in agreement with eqn (8.36), but $\text{Var}(G_C) = 2(1 - r)$ which is always greater than $2/3$. Knowledge of the genotypic values of the grandparents has affected the residual error variance for two reasons. Firstly it has contributed further information about the number of heterozygous loci in the two parents. Secondly genes at different loci in the child are no longer statistically independent if there is linkage; it is for this reason that the residual error variance depends on the recombination fraction.

Suppose now that A and C are two relatives of any kind with coefficient of relationship $R$. It is easy to show under the above model that the regression of A on C is linear:

$$E[G_A | G_C] = (1 - R)2np + RG_C. \qquad (8.38)$$

For it is clear that $G_{iA}$ is independent of $G_{jC}$ for $i \neq j$, and that the Expected value of $G_{iA}$ given $G_{iC}$ is $2p$, $p + 0.5G_{iC}$, or $G_{iC}$ according as the number of pairs of identical genes at the $i$th locus is 0, 1, or 2. However, it is not true that the regression of an individual on two or more relatives is necessarily linear under this model. As a counter-example suppose that A is the father of C who is the father of D. The mean square linear regression of C on A and D is

$$G_C = 0.4np + 0.4G_A + 0.4G_D.$$

Suppose now that $G_A = 0$ and $G_D = 2n$, so that A is homozygous for $b$ and $D$ is homozygous for $B$ at all loci. Then C must be heterozygous at all loci since at each locus he has inherited $b$ from his father and transmitted $B$ to his son. Hence

$$E(G_C | G_A = 0, G_D = 2n) = n,$$

and furthermore the residual error about this prediction is zero. The prediction from the mean square linear regression is $(0.4p + 0.8)n$, which is not in agreement with the true prediction unless $p = 0.5$. It can be concluded that the true regression is non-linear when $p \neq 0.5$.

The argument leading to the linear regressions (8.31), (8.36), and (8.38) can easily be adapted to any situation in which genes act additively. It can therefore be applied to the additive component of the genotypic value, $A$, under any genetic model, even when dominance and epistasis are present. In particular, the regression of the additive component of a child on the additive components of both parents is

$$E(A_C | A_F, A_M) = 0.5A_F + 0.5A_M \qquad (8.39)$$

This equation also expresses the regression on the additive components of both parents plus any other direct ancestors, and is valid whether or not there is linkage.

*Dominance*

We turn now to the effect of dominance on regressions between genotypic values. We modify the model by making $B$ completely dominant over $b$; we take $G_i = 0$ if the genotype at the $i$th locus is $bb$ and $G_i = 1$ otherwise.

Consider first the regression of a child on his father. If $G_{iF} = 1$, the chances that the father has 2 or 1 $B$ genes at the $i$th locus are $p/(1 + q)$ and $2q/(1 + q)$ respectively. Hence

$$\Pr(G_{iC} = 1 | G_{iF} = 1) = (p + q + pq)/(1 + q) = (1 + pq)/(1 + q). \quad (8.40)$$

If $G_{iF} = 0$ then

$$\Pr(G_{iC} = 1 | G_{iF} = 0) = p.$$

The conditional distribution of $G_C$ given $G_F$ is the sum (convolution) of two binomial distributions, the first with probability given by eqn (8.40) and index $G_F$ and the second with probability $p$ and index $n - G_F$. Hence

$$E(G_C | G_F) = np + q(1 + q)^{-1} G_F$$
$$\text{Var}(G_C | G_F) = npq + q(2q^2 - 1)(1 + q)^{-2} G_F. \quad (8.41)$$

It will be seen that the regression is linear despite dominance.

We now extend this argument to the biparental regression. We first observe that

$$\Pr(G_{iC} = 1 | G_{iF} = G_{iM} = 1) = 1 - q^2(1 + q)^{-2}$$
$$\Pr(G_{iC} = 1 | G_{iF} = 1, G_{iM} = 0 \text{ or vice versa}) = (1 + q)^{-1}$$
$$\Pr(G_{iC} = 1 | G_{iF} = G_{iM} = 0) = 0.$$

Denote by $n_j$ the number of loci at which $G_{iF} + G_{iM} = j (j = 0, 1, 2)$. Then

$$E(G_{iC} | n_0, n_1, n_2) = (1 + q)^{-1} n_1 + (1 - q^2(1 + q)^{-2}) n_2.$$

Conditional on $G_F$ and $G_M$,

$$E(n_2 | G_F, G_M) = G_F G_M / n$$
$$E(n_1 | G_F, G_M) = (G_F(n - G_M) + G_M(n - G_F))/n = (G_F + G_M) - 2G_F G_M / n.$$

(Note: the probability that $G_{iF} = G_{iM} = 1$ is $(G_F/n) \times (G_M/n)$.) Hence

$$E(G_C | G_F, G_M) = (1 + q)^{-1} (G_F + G_M) - (1 + q)^{-2} n^{-1} G_F G_M. \quad (8.42)$$

This is a hyperbola, but tends to a straight line when $n$ is large. This result was obtained by Karl Pearson (1904).

Equation (8.41) shows that the regression of child on parent is linear under a model with complete dominance. The argument can be extended to show that the regression between any pair of relatives is linear under this model. The result is rather misleading, however, since it ceases to be true when dominance is partial. Consider for simplicity a single locus model, the genotypes $bb$, $Bb$, and $BB$ having values 0, $d$, and 1. A straightforward calculation gives the regression of child on parent as:

$$G_F \qquad 0 \qquad d \qquad 1$$
$$E(G_C|G_F) \quad pd \quad 0.5d + 0.5p \quad qd + p$$

This regression is clearly non-linear unless $d = 0.5$ (no dominance), since there is a constant difference between successive values of $E(G_C|G_F)$.

To assess the degree of non-linearity in the offspring–parent regression due to dominance, consider a character determined by $n$ loci each with the same effect and with the same gene frequency, the genotypes $bb$, $Bb$, and $BB$ contributing 0, $d$, and 1 to the genotypic value. The exact regression of child on parent is difficult to find, so we shall instead compute the mean square quadratic regression of the form

$$G_C = a_0 + a_1 G_F + a_2 G_F^2 + e \tag{8.43}$$

which minimizes $E(e^2)$. This regression is more conveniently expressed in terms of orthogonal polynomials

$$G_C = b_0 \xi_0 + b_1 \xi_1 + b_2 \xi_2 + e \tag{8.44}$$

with

$$\xi_0 = 1$$
$$\xi_1 = (G_F - m)$$
$$\xi_2 = (G_F - m)^2 - \frac{m_3}{m_2}(G_F - m) - m_2 \tag{8.45}$$

where $m$ is the mean value of $G$, and $m_2$ and $m_3$ are the second and third moments about the mean. (See Draper and Smith (1966) for a discussion of orthogonal polynomials.)

As a result of orthogonality we have the simple results

$$b_r = E(G_C \xi_r)/E(\xi_r^2)$$
$$R(b_r) = E^2(G_C \xi_r)/E(\xi_r^2) \tag{8.46}$$

where $R(b_r)$ denotes the variance removed by fitting $b_r$. Hence we find that $b_0 = m$ and $b_1 = \frac{1}{2}V_A/V_G$, as expected. The quadratic component is most meaningfully expressed in terms of the moments of $G_i$, the single locus contributions, defined as

$$\mu = E(G_i)$$
$$\mu_r = E(G_i - \mu)^r \quad\quad (8.47)$$
$$\mu_{1,r} = E[(G_{iC} - \mu)(G_{iF} - \mu)^r].$$

We find that

$$b_2 = (\mu_2 \mu_{1,2} - \mu_3 \mu_{1,1})/2n\,\mu_2^3 + o\!\left(\frac{1}{n}\right) \quad\quad (8.48)$$

$$R(b_2)/R(b_1) = (\mu_2 \mu_{1,2} - \mu_3 \mu_{1,1})^2/2n\,\mu_{1,1}^2\,\mu_2^3 + o\!\left(\frac{1}{n}\right).$$

Some numerical results are shown in Table 8.4. It can be concluded that the only important contributions to non-linearity are likely to come from rather rare, almost recessive genes.

TABLE 8.4

*Non-linearity of parent–offspring regression under dominance*

| $d$ | $p$ | Sign of $b_2$ | $nR(b_2)/R(b_1)$ |
|---|---|---|---|
| 0 | 0.1 | 0 | 0 |
| 0.1 | 0.1 | − | 3.926 |
| 0.25 | 0.1 | − | 0.529 |
| 0.5 | 0.1 | 0 | 0 |
| 0.75 | 0.1 | + | 0.001 |
| 0.9 | 0.1 | + | 0.000 |
| 1.0 | 0.1 | 0 | 0 |
| 0 | 0.5 | 0 | 0 |
| 0.1 | 0.5 | − | 0.009 |
| 0.25 | 0.5 | − | 0.025 |
| 0.5 | 0.5 | 0 | 0 |
| 0.75 | 0.5 | + | 0.025 |
| 0.9 | 0.5 | + | 0.009 |
| 1.0 | 0.5 | 0 | 0 |

*Phenotypic regressions*

We have so far considered regressions between genotypic values, in particular the regression of $G_C$ on $G_F$. It is natural to assume that if this regression is linear then the regression between phenotypic values will also be linear, but this is not necessarily the case. We first observe that

$$E(Y_C|Y_F) = E(G_C + E_C|Y_F) = E(G_C|Y_F),$$

since $E_C$ is distributed independently of $Y_F$. Thus we need only consider the regression of $G_C$ on $Y_F$, though the variance about the regression will of course be increased by $V_E$. It will be assumed that the regression of $G_C$ on $G_F$ is linear.

We shall now compute the mean square quadratic regression of $G_C$ on $Y_F$:

$$G_C = a_0 + a_1 Y_F + a_2 Y_F^2 + e = b_0 \xi_0 + b_1 \xi_1 + b_2 \xi_2 + e, \qquad (8.49)$$

with the $\xi_r$s defined as before but with phenotypic values and moments. We find that $b_0 = m$ and $b_1 = \frac{1}{2} V_A / V_Y$ as expected. To find $b_2$ we compute

$$E(G_C \, \xi_2) = \frac{1}{2} \frac{V_A}{V_G} (m_2^* m_3 - m_2 m_3^*)/V_Y \qquad (8.50)$$

$$E(\xi_2^2) = m_4 + m_4^* + 6m_2 m_2^* + (m_3 + m_3^*)^2/V_Y - V_Y^2,$$

where $m_r$ and $m_r^*$ are the moments of $G$ and $E$ respectively. (To compute $E(G_C \xi_2)$ we use the result that $E[(G_C - m)(G_F - m)^2] = \frac{1}{2} V m_3 / V_G$. This follows from multiplying both sides of the regression of $G_C$ on $G_F$ by $(G_F m)^2$ and taking Expected values.) Thus the condition for $b_2$ to be zero is that

$$m_3^*/m_3 = m_2^*/m_2. \qquad (8.51)$$

This result is a special case of a general theorem due to Lindley (1947). Suppose that $X_1$ and $X_2$ are random variables and that the regression of $X_2$ on $X_1$ is linear. If $Y_1 = X_1 + E_1$ and $Y_2 = X_2 + E_2$, where $E_1$ and $E_2$ are errors of measurement distributed independently of the $X$s, the condition that the regression of $Y_2$ on $Y_1$ is linear is that the cumulants of $E_1$ are proportional to the cumulants of $X_1$ so that

$$\kappa_r(E_1) = \frac{\kappa_2(E_1)}{\kappa_2(X_1)} \kappa_r(X_1), \quad \text{for all } r \geqslant 3. \qquad (8.52)$$

For example, if $E_1$ is normal then $X_1$ must also be normal to meet this condition.

To obtain an idea of the amount of non-linearity which can arise in this way, suppose that $G$ and $E$ have the same variance, that they have third moments equal in size but opposite in sign, and that they are approximately normal so that normality can be assumed in computing $E(\xi_2^2)$. We find that

$$b_2 = m_3/16m_2^2$$

$$R(b_2)/R(b_1) = \frac{1}{4} m_3^2/m_2^3. \qquad (8.58)$$

For further discussion see Robertson (1977).

*Linkage*

In the absence of epistasis linkage has no effect on regressions between relatives. Consider a pair of relatives with genotypic values $G'$ and $G''$. The regression of $G''$ on $G'$ is

$$E(G''|G') = \sum_i E(G_i''|G'). \tag{8.54}$$

The distribution of $G_i''$ only depends on $G'$ through the information contained in $G'$ about $G_i'$; furthermore the distribution of $G_i'$ given $G'$ is unaffected by linkage provided that the population is in linkage equilibrium. Hence the distribution of $G_i''$ given $G'$, and in particular its mean, are unaffected by linkage, and will thus become linear when the number of loci is large regardless of their linkage relationships. This argument can be extended to the regression of an individual on two or more relatives, and to regressions involving additive effects and dominance deviations.

The effect of linkage is not on the regression but on the distribution of the residual error about the regression, and in particular its variance. The residual error variance is

$$\mathrm{Var}(G''|G') = \sum_i \mathrm{Var}(G_i''|G') + \sum_{i \neq j} \mathrm{Cov}(G_i'', G_j''|G'). \tag{8.55}$$

The conditional variance of $G_i''$ is unaffected by linkage, but the conditional covariance of $G_i''$ and $G_j''$, which is zero for loci on different chromosomes, may become non-zero for linked loci.

To illustrate the effect of linkage on the variance we shall consider an extreme situation with virtually complete linkage between all pairs of loci but with linkage equilibrium; no assumptions are made about the genetic model except that there is a large number of loci without epistasis. With complete linkage the number of identical genes must be the same at all loci, so that the conditional distribution of $G''$ given $G'$ is a mixture of three normal distributions: (i) with probability $P_0$ there are no identical genes at any locus and $G''$ will be normal with mean $m$ and variance $V_G$; (ii) with probability $P_1$ there is one identical gene at every locus, so that the distribution of $G''$ given $G'$ will be the same as that of child given parent and will be normal with mean $(1 - \frac{1}{2}h^2)m + \frac{1}{2}h^2 G'$ and with variance $(1 - \frac{1}{4}h^4)V_G$, where $h^2 = V_A/V_G$; (iii) with probability $P_2$ there are two identical genes at every locus so that $G'' = G'$. The regression of $G''$ on $G'$ is linear with slope $(\frac{1}{2}h^2 P_1 + P_2)$ as in the absence of linkage. The variance about the regression line is

$$\mathrm{Var}(G''|G') = [1 - (\tfrac{1}{2}h^2 P_1 + P_2)^2]V_G$$
$$+ [\tfrac{1}{4}h^4 P_1 + P_2 - (\tfrac{1}{2}h^2 P_1 + P_2)^2] \times [(G' - m)^2 - V_G]. \tag{8.56}$$

The argument can be repeated with phenotypic values, and it will be found that

$$\mathrm{Var}(Y''|Y') = [1 - (\tfrac{1}{2}h^2 P_1 + h_w^2 P_2)^2]V_Y$$
$$+ [\tfrac{1}{4}h^4 P_1 + h_w^4 P_2 - (\tfrac{1}{2}h^2 P_1 + h_w^2 P_2)^2] [(Y' - m)^2 - V_Y], \tag{8.57}$$

where $h^2 = V_A/V_Y$ and $h_w^2 = V_G/V_Y$.

In the absence of linkage the variance about the regression line is given by the first term on the right hand side of eqns (8.56) or (8.57). The effect of the second term is to decrease the variance when $G'$ or $Y'$ is near the mean and to increase it away from the mean. Note that this term is zero for parent and child, as predicted. For sibs, by contrast, we find when there are equal amounts of additive genetic and environmental variance and no dominance that

$$\text{Var}(G''|G') = \tfrac{5}{8}V_G + \tfrac{1}{8}(G' - m)^2$$
$$\text{Var}(Y''|Y') = \tfrac{29}{32}V_Y + \tfrac{1}{32}(Y' - m)^2. \tag{8.58}$$

In the more realistic case with partial linkage, a solution has been found (Bulmer 1976a) in the situation with two alleles with equal and additive effects and with the same gene frequency at each locus. With a large number of loci

$$\text{Var}(G''|G') = (1 - \bar{S})V_G + (\bar{S} - R^2)(G' - m)^2$$
$$\text{Var}(Y''|Y') = (1 - h^4 \bar{S})V_Y + (\bar{S} - R^2)h^4 (Y' - m)^2. \tag{8.59}$$

In these equations $R = (\tfrac{1}{2}P_1 + P_2)$, and $S = (\tfrac{1}{4}P(1, 1) + \tfrac{1}{2}P(1, 2) + \tfrac{1}{2}P(2,1) + P(2, 2))$, where $P(i, j)$ is the probability that there are $i$ pairs of identical genes at one locus and $j$ pairs at another linked locus; $S$ will depend on the recombination fraction between the loci, and $\bar{S}$ is its average value over all pairs of loci.

In the absence of linkage $\bar{S} = R^2$ so that the variance is constant, but in the presence of linkage $\bar{S} > R^2$ for any relatives except parent and child so that the variance about the regression line increases as $G'$ moves away from its mean value. To evaluate the magnitude of this effect it is necessary to estimate a typical value for $(\bar{S} - R^2)$. For sibs it follows from the answer to Problem 9, Chapter 3, that $S - R^2 = \tfrac{1}{8} - \tfrac{1}{4}r(1 - r)$ for a pair of loci with recombination fraction $r$. If genes are distributed at random on the chromosome map in a species with $c$ chromosomes each of length 1 morgan, it can be shown by using the usual mapping function that $(\bar{S} - R^2) = 0.047/c$ for sibs. This effect will be too small to be experimentally detectable, and may also be counterbalanced by the effect of having only a finite number of loci. Nevertheless it may have an appreciable cumulative effect when selection is continued over several generations, as we shall see in Chapter 9.

*Assortative mating*

In the last section we considered the effects of assortative mating under the assumptions underlying the ideal normal theory (an effectively infinite number of unlinked loci without epistasis). To gain some insight into the likely effects of departures from normal theory we shall now try to obtain some approximate results for a model with a finite number of unlinked loci with additive gene action. We write

$$G = X_{p1} + X_{m1} + X_{p2} + \cdots + X_{pn} + X_{mn}, \tag{8.60}$$

where $X_{pi}$ and $X_{mi}$ are respectively the contributions from the paternal and maternal genes at the $i$th locus. Define the variances and covariances of these contributions as

$$C_{ij} = \text{Cov}(X_{pi}, X_{pj}) = \text{Cov}(X_{mi}, X_{mj})$$
$$C'_{ij} = \text{Cov}(X_{pi}, X_{mj}), \tag{8.61}$$

and write

$$R_i(t) = \sum_j (C_{ij}(t) + C'_{ij}(t)). \tag{8.62}$$

The genetic variance is

$$V_G(t) = 2 \sum_i R_i(t); \tag{8.63}$$

the genetic variance under random mating is

$$V_G \equiv V_G(0) = 2 \sum_i C_{ii}(0). \tag{8.64}$$

We shall now develop some recurrence relationships for these variances and covariances under assortative mating. Since there is no selection the distribution of genes at different loci inherited from the same parent is unaffected by assortative mating, so that

$$C_{ij}(t + 1) = (1 - r_{ij})C_{ij}(t) + r_{ij} C'_{ij}(t). \tag{8.65}$$

In the steady state

$$r_{ij}(\hat{C}_{ij} - \hat{C}'_{ij}) = 0$$

so that

$$\hat{C}_{ij} = \hat{C}'_{ij} \quad \text{for } i \neq j. \tag{8.66}$$

We now observe that $C'_{ij}(t + 1)$ is equal to the covariance between genic effects in mates in generation $t$. Let $X^F_{pi}$ and $X^M_{pi}$ denote typical effects in father and mother in generation $t$. The mean square linear regression of these effects on the corresponding phenotypic values are

$$X^F_{pi} = \frac{R_i(t)}{V_Y(t)} Y_F + e_F$$

$$\tag{8.67}$$

$$X^M_{pi} = \frac{R_i(t)}{V_Y(t)} Y_M + e_M$$

(We may assume without loss of generality that the mean is zero.) We now suppose as an approximation that these are the true regressions, and we observe that $e_F$ and $e_M$ are independently distributed when assortative mating acts through phenotypic values. Hence

$$C'_{ij}(t + 1) = \text{Cov}(X^F_{pi}, X^M_{pj}) = \frac{R_i(t) R_j(t)}{V_Y(t)} \rho. \tag{8.68}$$

In the steady-state

$$\hat{C}'_{ij} = \rho \hat{R}_i \hat{R}_j / \hat{V}_Y \tag{8.69}$$

From eqns (8.66) and (8.69)

$$\hat{V}_G = 2 \sum_i \hat{C}_{ii} + 2 \sum_{i \neq j} \hat{C}_{ij} + 2 \sum_{i,j} \hat{C}'_{ij}$$

$$= V_G + \frac{2\rho}{\hat{V}_Y} \sum_{i \neq j} \hat{R}_i \hat{R}_j + \frac{2\rho}{\hat{V}_Y} \sum_{i,j} \hat{R}_i \hat{R}_j$$

$$= V_G + \rho \frac{\hat{V}_G^2}{\hat{V}_Y} - 2\rho \frac{\sum \hat{R}_i^2}{\hat{V}_Y}. \tag{8.70}$$

When the number of loci is large the last term in eqn (8.70) will be small compared with the preceding term. If we ignore this term we obtain the normal theory result (8.21). If we suppose that the loci have equal effects, so that $\hat{R}_i$ is the same for all $i$, eqn (8.70) can be written

$$\hat{V}_G \left[ 1 - \rho \left( 1 - \frac{1}{2n} \right) \hat{h}^2 \right] = V_G. \tag{8.71}$$

This is the result which would obtain under normal theory with marital correlation $\rho(1 - 1/2n)$ instead of $\rho$. It will also be seen from eqn (8.70) that the steady-state change in the expressed genetic variability under assortative mating is independent of the linkage relations between the loci.

We turn now to the correlations between relatives. The results of the previous section depend, in the absence of dominance, on the linearity of three regressions: that of $A_C$ on $A_F$ and $A_M$ in eqn (8.15), that of $A_F$ on $Y_F$ in eqn (8.19), and that of $A_M$ on $Y_F$. It has been shown in this section that the first regression remains linear but that the second and third ones do not necessarily do so when the assumptions of ideal normal theory are relaxed. Thus the results can only be regarded as approximations in the general case, depending on the magnitude of departures from linearity in these regressions. It seems likely, however, that they will provide satisfactory approximations in most cases. It also seems likely, from similar considerations, that the ideal normal theory will give a satisfactory approximation in the presence of dominance, with the possible refinement that $(1 - 1/2n)\rho$ be substituted for $\rho$ in eqn (8.21).

## Problems

1. Let $Y_C$ be the phenotypic value of a child, $Y_P$ be the average value of his parents, and $Y_G$ the average of his four grandparents. Under multivariate normality find the regression of $Y_C$ on $Y_P$ and $Y_G$, and the residual variance about the regression.

2. A character has a mean value of 100, standard deviation 10, and heritability $\frac{1}{2}$. If an individual's parents have values 105 and 110, what is the chance that his own value will exceed 120? Assume multivariate normality.

3. A character has phenotypic variance 100, the correlation between mates is 0.4, and the regression of child on midparent is 0.6. If the population mates at random for several generations what will the phenotypic variance and the regression of child on mid-parent become, if the character is controlled (a) by a large number of loci, (b) by two loci?

4. Draw the pedigrees of some (or all) of the relationships in Table 8.2 and verify the covariances.

5. Let $G_C$ be the genotypic value of a child and $G_P$ be the average genotypic value of his parents, $G_P = \frac{1}{2}(G_F + G_M)$. Use eqns (8.36) and (8.37) to find the regression of $G_C$ on $G_P$ and the residual variance about this regression. (Note that $(G_F + G_M)^2 - (G_F - G_M)^2 = 4G_F G_M$, and that $(G_F + G_M)$ and $(G_F - G_M)$ are independently distributed if $G_F$ and $G_M$ have independent normal distributions, which may be regarded as a satisfactory approximation.)

6. Use eqn (8.42) to find the regression of $G_C$ on $G_P$ under a model with complete dominance.

7. Consider a trait determined by two loci each with two alleles; at each locus the genotypes $BB$, $Bb$, and $bb$ contribute 1, $d$, and 0 to the character and the frequency of $B$ is $p$. Find $E(G_C|G_F)$ for the six possible values of $G_F$.

8. Suppose that genotypic values are multinormal, that $E = X - \mu$ where $X$ is exponential with mean $\mu$ and that the heritability is $\frac{1}{2}$, all the genetic variance being additive. Find the mean square quadratic regression of $Y_C$ on $Y_F$ when $E(Y) = 0$, $Var(Y) = 1$.

# 9 Selection

*In Chapters 6 and 8 we have developed the theory of quantitative genetics in a random mating population in the absence of any disturbing factors such as genetic drift, mutation, migration, or selection. In this chapter we shall begin to investigate the effect of selection on a quantitative character in an outbred population.*

*There are two types of selection, artificial selection due to the action of the plant or animal breeder in choosing the parents of the next generation, and natural selection due to differential mortality or fertility in the population. Their common feature is that the parents of the next generation are a selected subgroup of the whole population. The problem of interest is to predict the effect of selecting the parents on the characteristics of the offspring and of future generations.*

*The effect of one generation of selection can be inferred from the observed regression of offspring on parent, without using any genetic theory, but it is necessary to appeal to genetic theory to predict the effect of selection over several generations since the regression may change under selection. Two types of change may occur: a permanent change due to changes in gene frequencies, and a temporary change, which is reversed when selection ceases, due to the build-up of linkage disequilibrium. In this chapter it will be assumed that the time scale is short enough that changes in gene frequencies can be ignored. We shall first describe the empirical theory of selection, valid for one generation only and based on the observed parent–offspring regression; we shall then use the normal distribution theory derived in Chapter 8 to investigate changes resulting from linkage disequilibrium; we shall finally discuss some of the consequences of departures from this ideal theory. The effect of selection on gene frequencies will be considered in the next chapter.*

*We shall restrict our attention to selection in an outbreeding species. The plant breeder working with a self-fertilizing species may wish to predict the effect of selection combined with selfing following a cross between two pure lines, but this theory is rather complicated and has not been extensively developed. The reader is referred to Mayo (1980) for a general account, and in particular to the work of Pederson (1969).*

## Empirical selection theory

In this section we shall consider the effect of one generation of selection based on the phenotypic values of the parents on the mean phenotypic value among their offspring. Suppose that the biparental regression has been determined empirically and has been found to be linear so that

$$E(Y_C | Y_F, Y_M) = a + b_F Y_F + b_M Y_M. \qquad (9.1)$$

The separate coefficients $b_F$ and $b_M$ allow for the possibility of sex differences; if they are present a separate regression must be fitted for male and female children. Write $m^*$ and $m^{**}$ for the mean phenotypic values in males and females in the absence of selection. Suppose now that parents are selected in some way, either by nature or by the breeder, with the result that the mean phenotypic values among fathers and mothers are $\overline{Y}_F$ and $\overline{Y}_M$ respectively. The Expected phenotypic value in their offspring is

$$E(Y_C | \overline{Y}_F, \overline{Y}_M) = a + b_F \overline{Y}_F + b_M \overline{Y}_M, \tag{9.2}$$

compared with a predicted value

$$a + b_F m^* + b_M m^{**} \tag{9.3}$$

in the absence of selection; the latter expression is equal to $m^*$ or $m^{**}$ depending on whether the regression for male or female children is being used.

We now write $S_F = (\overline{Y}_F - m^*)$ and $S_M = (\overline{Y}_M - m^{**})$; these quantities are the *selection differentials* among fathers and mothers, and they represent the change in the mean value among parents due to selection. The change in the Expected phenotypic value in the offspring as a result of selection is the predicted response to selection, denoted by $R$. It follows from eqns (9.2) and (9.3) that

$$R = b_F S_F + b_M S_M. \tag{9.4}$$

(Remember that if there are sex differences $b_F$ and $b_M$ will depend on the sex of the child.) If there are no sex differences so that $b_F = b_M = b$, it is convenient to define $\beta = 2b$ and $S = \frac{1}{2}(S_F + S_M)$; $\beta$ is the regression of offspring on mid-parent and $S$ is the mean of the selection differentials in the two parents. In this case

$$R = \beta S. \tag{9.5}$$

The important results (9.4) and (9.5) depend only on the assumption of the linearity of the biparental regression (9.1) and do not depend on any genetic assumptions. Nor do they depend on the parents being mated to each other at random.

These results allow us to predict the response in the offspring generation once the selection differential among the parents is known. To describe the properties of a general scheme of selection it is therefore necessary to know the selection differential which is likely to arise from it. As an example we shall consider truncation selection, the most important form of artificial selection, and we shall derive the Expected selection differential arising from it.

*Truncation selection*

Under truncation selection the individuals with the largest phenotypic values are selected as parents of the next generation and the rest are discarded. The selection differential depends on the proportion of the population which is selected; the smaller the proportion the larger the selection differential. The proportion selected may differ in the two sexes. (In animal breeding a much higher proportion of females than males must be used to maintain a constant population size; Pirchner (1969) states that in cattle about two-thirds of all female calves must be used for breeding, but only 5 per cent of all bull calves need be retained for breeding under natural service, and only 1 per cent with artificial insemination.)

To formalize this scheme we suppose that $n$ individuals of a particular sex have been measured and arranged in rank order; denote the phenotypic value of the $r$th largest individual by $y_{(r)}$ $(y_{(1)} > y_{(2)} > \cdots > y_{(n)})$. The $k$ largest individuals are selected, so that the proportion selected is $p = k/n$. The selection differential is

$$S = \sum_{r=1}^{k} y_{(r)}/k - m. \qquad (9.6)$$

It is convenient to define the intensity of selection as the standardized quantity $i = S/\sigma$, where $\sigma$ is the standard deviation of the underlying distribution in the absence of selection. (The asterisks for males and females have been dropped since we are only considering individuals of one sex.) We write $I$ for the Expected value of $i$, $I = E(S)/\sigma$. Thus

$$i = \sum_{r=1}^{k} x_{(r)}/k$$
$$I = E(i) = \sum_{r=1}^{k} E(x_{(r)})/k \qquad (9.7)$$

where $x_{(r)}$ is the standardized order statistic, $x_{(r)} = (y_{(r)} - m)/\sigma$.

The predicted intensity of selection, $I$, can be found exactly if the Expected values of the order statistics of the underlying distribution are known. The Expected values of the order statistics for an underlying normal distribution have been extensively tabulated (Fisher and Yates 1953, Table 20; Biometrika Tables, Vol. 1, Table 28; Biometrika Tables, Vol. 2, Table 9). For large $n$, $I$ can be found approximately as the Expected value of a truncated normal distribution. Suppose that $z$ is the upper $100p$ per cent point of the standard normal distribution (so that $\Pr[Z \geqslant z] = p$), and let $X$ be a standard normal variate truncated at $z$ (values below $z$ being discarded). The density function of $X$ is $\phi(x)/p$ $(x \geqslant z)$, where

$$\phi(x) = (2\pi)^{-1/2} \exp -\tfrac{1}{2}x^2. \qquad (9.8)$$

The Expected value of $X$ is

$$E(X) = \int_{z}^{\infty} x\phi(x)\, dx/p. \qquad (9.9)$$

It is easily verified that

$$\phi'(x) = -x\phi(x), \qquad (9.10)$$

so that

$$E(X) = \phi(z)/p. \tag{9.11}$$

For large $n$ eqn (9.11) provides a good approximation to $I$, but for smaller values of $n$ it overestimates $I$. A better approximation is obtained by using eqn (9.11) with $p$ redefined as

$$p = (k + \tfrac{1}{2})/(n + k/2n), \tag{9.12}$$

and with $z$ redefined accordingly. When $k/n$ is small we may equivalently redefine $p$ as $(k + \tfrac{1}{2})/n$. Some exact values of $I$ are shown in Table 9.1. The maximum error in using eqn (9.11) as an approximation with $p$ redefined as in eqn (9.12) was two in the second decimal place.

TABLE 9.1

*Expected intensity of selection (I) under truncation selection from an under-lying normal distribution*

| Proportion selected | $n = 10$ | $n = 20$ | $n = 50$ | $n = 100$ | $n = \infty$ |
|---|---|---|---|---|---|
| 0.01 | | | | 2.51 | 2.67 |
| 0.02 | | | 2.25 | 2.33 | 2.42 |
| 0.03 | | | | 2.20 | 2.27 |
| 0.04 | | | 2.05 | 2.10 | 2.15 |
| 0.05 | | 1.87 | | 2.02 | 2.06 |
| 0.1 | 1.54 | 1.64 | 1.71 | 1.73 | 1.75 |
| 0.2 | 1.27 | 1.33 | 1.37 | 1.39 | 1.40 |
| 0.3 | 1.07 | 1.11 | 1.14 | 1.15 | 1.16 |
| 0.4 | 0.89 | 0.93 | 0.95 | 0.96 | 0.97 |
| 0.5 | 0.74 | 0.77 | 0.79 | 0.79 | 0.80 |
| 0.6 | 0.60 | 0.62 | 0.63 | 0.64 | 0.64 |
| 0.7 | 0.46 | 0.48 | 0.49 | 0.49 | 0.50 |
| 0.8 | 0.32 | 0.33 | 0.34 | 0.35 | 0.35 |
| 0.9 | 0.17 | 0.18 | 0.19 | 0.19 | 0.19 |
| 1.0 | 0 | 0 | 0 | 0 | 0 |

**Normal distribution theory**

We saw in the last section how the effect of one generation of selection can be derived by considering the regression of offspring on their parents, which can be estimated empirically. To find the effect of a further generation of selection it is necessary to consider the regression on both their parents and their grandparents. In general it will be necessary to consider the regression of an individual on all his direct ancestors back to the generation when selection began. It is clearly impracticable to treat this problem empirically, without some theoretical basis for simplifying these regressions. In this section we shall investigate the effect of selection over several generations under the assumptions underlying the normal distribution theory described in Chapter 8.

Consider a character for which multivariate normality holds in a population at equilibrium under random mating with no selection. Suppose now that some form of selection comes into operation, acting on genotypic or phenotypic values in the population for one or more generations. The distribution of dominance deviations, $D$, in any generation, measured before selection, is unaffected by the action of selection in previous generations. The reason is that without selection $D$ is distributed independently of ancestral values. We may therefore condition on ancestral values in any way, or we may select ancestral values in any way (which comes to the same thing), without changing the distribution of $D$ in the present generation. Thus $D$, measured before selection, remains distributed normally and independently of $A$ with zero mean and variance $V_D$. Animal breeders usually refer to the additive component, $A$, as the *breeding value* and we shall treat these terms as synonymous.

The distribution of the breeding value, $A$, is of course affected by selection but the regression of the breeding value of an individual, measured before selection, on the breeding values of his parents remains invariant. Under random mating with no selection this regression is

$$A_C = \tfrac{1}{2}A_F + \tfrac{1}{2}A_M + e, \tag{9.13}$$

where the residual error, $e$, is normal with zero mean and variance $\tfrac{1}{2}V_A$, independently of the breeding values or dominance deviations of any direct ancestors of C. By the same argument as before it follows that the distribution of $A_C - \tfrac{1}{2}(A_F + A_M)$, measured before selection in the generation of the offspring C, is unaffected by any form of selection in generations before that of C; it is thus normal with zero mean and variance $\tfrac{1}{2}V_A$, independently of the breeding values or dominance deviations of the parents or any other ancestors of C, where $V_A$ is the variance of the breeding values in a population mating at random without selection. Thus the regression (9.13) is unaffected by selection.

Eqn (9.13) is the basic result from which the effect of selection on succeeding generations can be deduced. If we know the distribution of $A$ in the parental generation after selection, that is to say the joint distribution of $A_F$ and $A_M$, we can find the distribution of $A$ in the next generation before selection (the distribution of $A_C$) by the standard methods for finding the distribution of sums of random variables. Hence the distribution of breeding values and of phenotypic values can be found recursively in successive generations.

In particular suppose that selection acts equally on the two sexes and that mating is random so that $A_F$ and $A_M$ are independently and identically distributed. Then the cumulant generating function of $A$ satisfies the relationship

$$K_{t+1}(z) = 2K_t^*(\tfrac{1}{2}z) + \tfrac{1}{4}V_A z^2, \tag{9.14}$$

where $K_t^*(z)$ is the c.g.f. of $A$ in generation $t$ *after* selection and $K_{t+1}(z)$ is the c.g.f. in the next generation *before* selection. For the mean we find

$$m_A(t + 1) = m_A^*(t). \tag{9.15}$$

Thus any change in the mean breeding value among the selected group of parents is transmitted in its entirety to their offspring. For the variance

$$V_A(t + 1) = \tfrac{1}{2}V_A^*(t) + \tfrac{1}{2}V_A. \tag{9.16}$$

If we define $\delta = V_A(t) - V_A$ as the deviation of the additive variance from its equilibrium value in the absence of selection, then

$$\delta(t + 1) = \tfrac{1}{2}\delta^*(t). \tag{9.17}$$

Thus only half of the deviation in variance in the selected group of parents is transmitted to their offspring. For the higher cumulants

$$k_r(t + 1) = \tfrac{1}{2}^{r-1} k_r^*(t), \quad r \geqslant 3. \tag{9.18}$$

(The reader who is unfamiliar with cumulants and their generating functions should be able to derive the important results (9.15) and (9.16) directly from eqn (9.13).)

It is of interest to see what happens when selection is relaxed. Suppose that selection ceases in generation $T$ so that thereafter starred and unstarred quantities are the same. From the results of the previous paragraph we find that $t$ generations after selection was relaxed

$$m(T + t) = m(T)$$
$$\delta(T + t) = \tfrac{1}{2}^t \, \delta(T) \tag{9.19}$$
$$k_r(T + t) = \tfrac{1}{2}^{(r-1)t} k_r(T), \quad r \geqslant 3.$$

Thus any change in the mean brought about by selection is permanent, but the variance and higher moments quickly return to their original values when selection is relaxed. The cumulant generating function satisfies

$$K_{T+t}(z) = 2^t K_T(\tfrac{1}{2}^t z) + \tfrac{1}{2}(1 - \tfrac{1}{2}^t)V_A z^2. \tag{9.20}$$

As $t \to \infty$, the c.g.f. tends to the limit

$$m(T)z + \tfrac{1}{2}V_A z^2, \tag{9.21}$$

which is the c.g.f. of a normal variate with mean $m(T)$ and variance $V_A$. Thus the only permanent effect of selection is to shift the mean of the distribution of breeding values.

These results can be interpreted genetically as follows. Selection acts in two ways, by changing gene frequencies and by inducing departures from Hardy–Weinberg and linkage equilibrium. When selection is relaxed the effect of any changes in gene frequency will remain permanently but the effect of departures from linkage equilibrium will gradually disappear; Hardy–Weinberg equilibrium is restored immediately in each generation before selection. In the absence of epistatic interactions, linkage disequilibrium has no effect on the mean, so that any changes in the mean, measured before selection, must be due to changes in gene frequencies and will be permanent. On the other hand, since the model presupposes an effectively infinite number of loci with infinitesimal effects, a finite change in the mean can be brought about by an infinitesimal change in each of the gene frequencies which will have a negligible effect on the variance and the higher moments in the absence of linkage disequilibrium. Thus any change in the variance and the higher moments is due to linkage disequilibrium and will disappear when selection is relaxed. In particular the component of the variance $\delta$ representing the deviation from its equilibrium value is due entirely to pairwise correlations between loci. This quantity is halved from one generation to the next as shown in eqn (9.17) because the coefficient of linkage disequilibrium between unlinked loci is halved.

These results should be contrasted with the effect of selection on a mixture of pure lines. In this case, if the genotypic value after selection has some distribution $f^*$, of whatever shape, this distribution will be reproduced exactly in future generations in the absence of selection because pure lines breed true.

It is important to remember that normal theory presupposes an effectively infinite number of unlinked loci with infinitesimal effects; in consequence a finite change in the mean can be brought about by an infinitesimal change in gene frequencies. Thus normal theory ignores any changes in gene frequency which must occur in any real situation with a finite number of loci having finite effects. It is for this reason that, under normal theory, all changes in the variance and higher moments disappear when selection is relaxed.

To illustrate this point consider a model with $n$ loci, at each of which there are two alleles which contribute $a$ and $b$ respectively to the character, having frequencies $q$ and $p$. The genetic variance in gametic equilibrium is $2npq(b-a)^2$. If we let $n$ increase while keeping the genetic variance constant we must regard $(b-a)$ as a quantity of order $n^{-1/2}$. An increase in the mean by a fixed amount $d$ is brought about by an increase in all the gene frequencies by an amount $d/2n(b-a)$, which is of order $n^{-1/2}$. Thus a finite change in the mean is brought about by an infinitesimal change in all of the gene frequencies. Another way to make this point is to say that a fixed selection pressure acting on phenotypic values exerts a pressure of order $n^{-1/2}$ at each locus; we shall consider the selective force acting at individual loci in more detail in the next chapter.

We shall now consider in turn some different types of selection, starting with selection for an optimal value.

*Optimizing selection*

Suppose that the fitness of an individual with phenotypic value $y$ is

$$w(y) = \exp - \tfrac{1}{2}(y - \theta)^2/\gamma. \tag{9.22}$$

This model was introduced by Haldane (1954) to represent natural selection for an optimal value $\theta$. The parameter $\gamma$, which is analogous to the variance of a normal distribution, can be interpreted as the spread of fitness about the optimal value; the smaller $\gamma$, the greater is the intensity of selection. The fitness $w(y)$ can be thought of as the relative pre-reproductive mortality as a function of $y$. If the phenotypic value *before* selection has density function $f(y)$, its density function *after* selection will be

$$f^*(y) = w(y)f(y)/\int w(y)f(y)\,dy.$$

If $f(y)$ is normal with mean $m(t)$ and variance $V(t)$ (as it will be under normal theory), then

$$f^*(y) \propto \exp -\tfrac{1}{2}(y-\theta)^2/\gamma \cdot \exp -\tfrac{1}{2}(y-m(t))^2/V(t)$$

$$\propto \exp -\tfrac{1}{2}\left[y - \frac{\theta V(t)+\gamma m(t)}{V(t)+\gamma}\right]^2 \Big/ \frac{\gamma V(t)}{V(t)+\gamma}.$$

Thus the distribution of $Y$ after selection is normal with mean and variance

$$m^*(t) = \frac{\theta V(t)+\gamma m(t)}{V(t)+\gamma} = m(t) + \frac{V(t)}{V(t)+\gamma}[\theta - m(t)]$$

$$(9.23)$$

$$V^*(t) = \frac{\gamma V(t)}{V(t)+\gamma} = V(t) - \frac{V^2(t)}{V(t)+\gamma}.$$

We must now consider the effect of the changes in the phenotypic mean and variance on the distribution of breeding values. Before selection assume that $A$, $D$, and $E$ are independently and normally distributed, so that the regression of $A$ on $Y$ is

$$A = (1 - h^2(t))\,m(t) + h^2(t)Y + e \qquad (9.24)$$

where $e$ is normal with zero mean and variance $(1 - h^2(t))V_A(t)$. Since $Y$ is normal after selection, then so is $A$, since it is a linear function of the independent normal variates $Y$ and $e$. By taking Expectations in eqn (9.24) the mean and variance of $A$ after selection are found to be

$$m_A^*(t) = m(t) + \frac{h^2(t)V(t)}{V(t)+\gamma}(\theta - m(t)),$$

$$(9.25)$$

$$V_A^*(t) = V_A(t) - V_A^2(t)/(V(t)+\gamma).$$

Since $A$ is normally distributed after selection, it will still be normal in the next generation before selection (from the biparental regression (9.13)), and the argument can be repeated recursively under continued selection.

From eqn (9.16) we find that the additive variance before selection obeys the recursion

$$V_A(t + 1) = \tfrac{1}{2} V_A(t) \frac{(\gamma + V_D + V_E)}{(V_A(t) + \gamma + V_D + V_E)} + \tfrac{1}{2} V_A. \tag{9.26}$$

Starting from $V_A$, $V_A(t)$ will decrease quickly to an equilibrium value $\hat{V}_A$, which can be found by putting $V_A(t + 1) = V_A(t) = \hat{V}_A$. This gives a quadratic equation for $\hat{V}_A$ with the unique positive solution:

$$\hat{V}_A = -\tfrac{1}{4}(\gamma + V_D + V_E - V_A) + \tfrac{1}{4}[(\gamma + V_D + V_E + V_A)^2 \\ + 8V_A(\gamma + V_D + V_E)]^{1/2}. \tag{9.27}$$

From eqn (9.15) the mean, measured before selection, obeys the recursion

$$m(t + 1) = m(t) + \frac{[\theta - m(t)]V_A(t)}{V_A(t) + \gamma + V_D + V_E}. \tag{9.28}$$

If $V_A(t)$ were constant $m(t)$ would converge geometrically to $\theta$, but the process is modified by the decline of $V_A(t)$ from $V_A$ to $\hat{V}_A$, which will bring about a corresponding decline in the rate of convergence. A numerical example is shown in Table 9.2.

TABLE 9.2

*Selection for an optimal value with*
$\theta = V_A = V_D + V_E = \gamma = 100, m(0) = 50$

| $t$ | $V_A(t)$ | $m(t)$ |
|---|---|---|
| 0 | 100 | 50 |
| 1 | 83 | 67 |
| 2 | 79 | 76 |
| 3 | 78 | 83 |
| 4 | 78 | 88 |
| 5 | 78 | 91 |
| Selection relaxed | | |
| 6 | 89 | 91 |
| 7 | 95 | 91 |
| 8 | 97 | 91 |
| 9 | 98 | 91 |
| 10 | 99 | 91 |

The special form of the fitness function for optimizing selection ensures that the breeding values remain normally distributed after selection. It is this fact which makes it possible to find a simple, exact solution. In other cases it is necessary to approximate to obtain a simple solution.

*Truncation selection*

We shall now consider *truncation selection* in which the individuals with the largest phenotypic values are selected as parents of the next generation and the rest are discarded. We shall suppose that in each generation a constant proportion, $p$, is selected from a very large population and we shall assume that this proportion is the same in both sexes. If the distribution of phenotypic values before selection is normal, the distribution among selected individuals will be a truncated normal distribution. The mean of this distribution was derived in eqn (9.9); we shall now obtain a recurrence relation for the higher moments.

Suppose that $z$ is the standard normal deviate corresponding to the proportion selected $p$, and let $X$ be a standard normal variate truncated at $z$, with density function $\phi(x)/p$, $x \geqslant z$. The $r$th absolute moment is

$$m'_r = \int\limits_z^\infty x^r \, \phi(x) \, \mathrm{d}x/p.$$

We observe that

$$\frac{\mathrm{d}}{\mathrm{d}x} x^r \phi(x) = rx^{r-1} \phi(x) + x^r \frac{\mathrm{d}\phi(x)}{\mathrm{d}x} = rx^{r-1} \phi(x) - x^{r+1} \phi(x).$$

Integrating both sides we find the recurrence relationship

$$m'_{r+1} = rm'_{r-1} + z^r \phi(z)/p.$$

Writing $c = \phi(z)/p$, we find that the mean, variance, and the third and fourth cumulants are

$$\begin{aligned}
m &= c \\
m_2 &= 1 - c(c - z) \\
k_3 &= c[(c - z)(2c - z) - 1] \\
k_4 &= c[-6c(c - z)^2 + (3 - z^2)(c - z) + c].
\end{aligned} \tag{9.29}$$

If the underlying distribution is normal with mean $m$ and variance $V$, the mean after truncation becomes $m + cV^{1/2}$ and the $r$th cumulant is multiplied by $V^{r/2}$. The variance is always reduced by truncation, and the skewness is of course positive (to the right).

In the first generation of selection the phenotypic values after selection will have a truncated normal distribution. From eqn (9.24) we can therefore find the moments of the distribution of breeding values after selection, and from eqns (9.15)–(9.18) we can find the moments in the next generation before selection. Unfortunately this argument cannot be repeated in subsequent generations since the distribution of breeding values is no longer normal. In consequence the phenotypic values after selection will not have a truncated normal distribution, and the regression of $A$ on $Y$ will no longer be linear and will no longer have constant residual error variance.

As an approximation we shall ignore departures from normality and proceed as if $A$ had a normal distribution. To justify this approach we shall first consider the effect of ignoring non-normality when all the variance is additive genetic variance. Suppose that initially $m = V_A = 100$, and that we select the top 5 per cent of the population. We can use normal theory to calculate that, in the next generation before selection, the moments will be

$$
\begin{aligned}
m &= 120.6 \\
V_A &= 56.9 \\
k_3 &= 18.9 \quad \text{skewness} = 0.0441 \\
k_4 &= 62.6 \quad \text{kurtosis} \; = 0.0193
\end{aligned}
$$

Approximating this distribution by a Gram–Charlier series (Kendall and Stuart 1963, Vol. 1, pp. 155–60), we find that truncation at the upper 5 per cent point will increase the mean to 136.4; the corresponding figure ignoring departures from normality is 136.2. The effect of non-normality will be even smaller when there is environmental or dominance variance to mask the breeding values. For example, when the initial heritability is $\frac{1}{2}$, the skewness and kurtosis of the breeding values before selection in the first generation after truncation selection starts with $p = 0.05$ are respectively only 0.0096 and 0.0025. It is concluded that departures from normality induced by selection can be safely ignored.

Assuming normality we can now find a recurrence relationship for the mean and variance under continued truncation selection:

$$
\begin{aligned}
m(t + 1) &= m(t) + ch(t)V_A^{1/2}(t) \\
V_A(t + 1) &= \tfrac{1}{2}[1 - h^2(t)c(c - z)]V_A(t) + \tfrac{1}{2}V_A.
\end{aligned}
\tag{9.30}
$$

Starting from $V_A$, $V_A(t)$ will decrease quickly to an equilibrium value $\hat{V}_A$ which can be found in the same way as with optimizing selection. The decrease in $V_A(t)$ will bring about a corresponding decrease in the rate at which the mean increases. A numerical example is shown in Table 9.3.

*General theory*

We now consider the effect of selection on the variance in a more general situation of which truncation selection is a special case. Suppose that departures from normality before selection can be ignored, and that the direct effect of selection on the phenotypic variance is to change it by a constant proportion: thus

$$
V^*(t) = \beta V(t),
\tag{9.31}
$$

TABLE 9.3

*Truncation selection with* p = 0.05 (c = 2.063, z = 1.645), *and with*
$$V_A = V_D + V_E = 100$$

| $t$ | $V_A(t)$ | $m(t + 1) - m(t)$ |
|---|---|---|
| 0 | 100 | 14.6 |
| 1 | 78 | 12.1 |
| 2 | 74 | 11.6 |
| 3 | 74 | 11.5 |
| 4 | 73 | 11.5 |
| 5 | 73 | 0 |
| | Selection relaxed | |
| 6 | 87 | 0 |
| 7 | 93 | 0 |
| 8 | 97 | 0 |
| 9 | 98 | 0 |
| 10 | 99 | 0 |

where $V(t)$ and $V^*(t)$ are the phenotypic variances before and after selection. From eqn (9.24)

$$V_A^*(t) = h^4(t)V^*(t) + (1 - h^2(t))V_A(t)$$
$$= [1 + (\beta - 1)h^2(t)]V_A(t). \tag{9.32}$$

From eqn (9.16),

$$V_A(t + 1) = \tfrac{1}{2}[1 + (\beta - 1)h^2(t)]V_A(t) + \tfrac{1}{2}V_A. \tag{9.33}$$

Starting from $V_A$, $V_A(t)$ will increase or decrease depending on whether $\beta$ is greater or less than 1 until it reaches a limiting value $\hat{V}_A$ which satisfies the quadratic equation

$$(2 - \beta)\hat{V}_A^2 + (V_D + V_E - V_A)\hat{V}_A - V_A(V_D + V_E) = 0. \tag{9.34}$$

When $\beta < 2$, there is a unique positive root towards which $V_A(t)$ will tend. When $\beta = 2$ there is a positive solution for $h^2 < \tfrac{1}{2}$, but not for $h^2 \geqslant \tfrac{1}{2}$; in the latter case $V_A(t)$ calculated from eqn (9.33) will increase without bound. When $\beta > 2$, there is a positive solution larger than $V_A$ if

$$h^2 < \tfrac{1}{2}$$

and

$$4h^2(1 - h^2) < 1/(\beta - 1). \tag{9.35}$$

If both these conditions are satisfied there are two positive roots larger than $V_A$; $V_A(t)$ will increase to the smaller of them. If they are not both satisfied, there may be complex roots, or real roots less than $V_A$ which cannot be reached; in these cases $V_A(t)$ calculated from eqn (9.33) will increase without limit.

Truncation selection is an example of this theory with $\beta = 1 - c(c - z)$. For optimizing selection $\beta = \gamma/(V + \gamma)$, which depends on the phenotypic variance. For both truncation and optimizing selection $\beta < 1$, so that the variance decreases. As an example of a type of selection which tends to increase the variance (often called disruptive selection) we shall consider double truncation in which the individuals with the smallest as well as the largest phenotypic values are chosen as parents, intermediate individuals being discarded; a fraction $p$ is chosen from each tail of the distribution, the total proportion selected being $2p$. As before we write $z$ for the standard normal deviate corresponding to $p$ and we write $c = \phi(z)/p$. The mean is unaltered by selection and the variance is increased from $V(t)$ to $(1 + cz)V(t)$, so that $\beta = (1 + cz)$. (This is the second moment about the origin of a singly truncated standard normal distribution.) $\beta$ exceeds the critical value of 2 when $p$ is less than 0.226. Thus when $p > 0.226$ the variance should increase to a finite upper limit which is the smaller of the two roots of eqn (9.34), while when $p < 0.226$ the variance should increase without limit.

### Simulation results

The main value of the theory developed in this section is its prediction of changes in the additive genetic variance under selection due to the build-up of linkage disequilibrium. The theory is based on the assumption that there is an infinite number of unlinked loci, and it is important to know how accurate the predictions are when the number of loci is moderate, say around ten. In a computer simulation to answer this question (see Bulmer 1976b for more detail) a character was assumed to be determined by 12 unlinked loci each with two alleles, + and −, contributing 1 and 0 respectively to the character without dominance or epistasis. The genotypic value was thus the number of + alleles, ranging from 0 to 24. The phenotypic value was obtained by adding a random normal deviate with zero mean and variance 4. The results of four experiments illustrating the effects of different types of selection on the variance are summarized in Table 9.4. The results for the first, third, and fourth experiments exclude the first ten generations during which quasi-equilibrium was attained, but the result for the second experiment (single truncation) uses only the first five generations since fixation occurred after 13 generations.

In all four experiments allowance must be made for the change in the gene frequencies due to the combination of selection and genetic drift. The genic variance, shown in the first column, is the genetic variance calculated from the observed gene frequencies on the assumption that the population is in perfect Hardy–Weinberg and linkage equilibrium:

$$\text{Genic variance} = 2\sum p_i q_i.$$

The second column shows the genotypic variance calculated from the observed genotypic values. The difference from the genic variance is due to correlations

TABLE 9.4

*Simulation of the effect of selection on the variance*

| Individuals selected | Genic variance | Genotypic variance (observed) | Genotypic variance (predicted) |
|---|---|---|---|
| Middle 20% | 4.75 | 3.29 | 3.29 |
| Top 20% | 3.82 | 2.97 | 2.88 |
| Bottom and top 10% | 6.00 | 70.53 | 72 |
| Bottom and top 25% | 5.73 | 16.57 | 18.37 |

between pairs of loci (linkage disequilibrium); apart from minor sampling fluctuations there is no correlation between the two alleles at the same locus (Hardy–Weinberg disequilibrium) under random mating. The third column shows the genotypic variance predicted from eqn (9.34) by substituting the genic variance for $V_A$, and $V_D = 0$, $V_E = 4$.

There is reasonably good agreement between the observed and predicted variances. The first experiment illustrates a type of selection for an optimal value; if the middle $100p$ per cent are selected it can be shown that $\beta = 1 - 2z\phi(z)/p$, where $z$ is the standard normal deviate such that $\Pr[-z \leqslant Z \leqslant z] = p$. The second experiment illustrates truncation selection; agreement with prediction is good in view of the rather rapid movement of the gene frequencies towards fixation which is ignored in the theory. It is also of interest that the 'realized heritability' in the first five generations was 0.43; this is equal to the heritability calculated from the genotypic variance, but is less than the heritability calculated from the genic variance, which is 0.49. (The 'realized heritability' is the ratio $R/S$, where $R$ is the response to selection and $S$ is the selection differential.) The third and fourth experiments illustrate truncation selection with proportions selected above and below the critical value of 0.226. For the third experiment the prediction is that $V_A(t)$ will increase without limit; the predicted value given in the table is the maximum variance attainable under random mating with the genotypic value taking the values 0, 12, and 24 with probabilities $\frac{1}{4}$, $\frac{1}{2}$, and $\frac{1}{4}$. The simulation attained this ideal state of complete linkage disequilibrium from generation 13 onwards, the only genotypic values observed being 0, 12, 24 in approximate ratios 1:2:1. The system thus mimicked a single locus with a gene frequency of $\frac{1}{2}$ with two alleles having values of 0 and 12.

**Departures from normal theory**

The results of the last section were obtained under the ideal conditions of normal distribution theory discussed on pages 123–6. The results of actual selection experiments may fail to agree with this ideal theory for a variety of reasons, of which the most important are the following:

(i) The parent–offspring regression may be non-linear, for reasons discussed on pages 131–40; in this case even the theory of page 145 will fail.

(ii) With a finite number of loci, selection will cause changes in gene frequencies which will lead to a permanent change in the genetic variance; under continued directional selection genetic variability will eventually become exhausted.

(iii) With small population size, random fluctuations due to genetic drift may mask the effect of selection.

(iv) In addition to the artificial selection imposed by the experimenter, the character may be subject to natural selection pressures unknown to him; their most likely effect will be to resist directional selection away from the normal value of the character, and to cause partial reversal of the mean value back to its starting point when artificial selection is relaxed.

(v) The presence of epistasis will also invalidate the proposition that any change in the mean under selection is permanent.

(vi) Linkage is likely to increase the degree of linkage disequilibrium induced by selection, and thus to exaggerate the temporary change in the additive genetic variance predicted in the last section in the absence of linkage.

We shall return to (ii) and (iii) in later chapters. In this section we will discuss the effects of linkage and epistasis on the response to selection.

### Linkage

The change in the variance under selection under normal theory is due entirely to the build-up of linkage disequilibrium, though it is rather inappropriately named since normal theory assumes all loci to be unlinked. One would expect a larger change in the variance in the presence of linkage, since the decay of disequilibrium through recombination will be slower. We shall now develop a rather heuristic theory which takes linkage into account.

Suppose for simplicity there is additive gene action, and write

$$G = X_{p1} + X_{m1} + X_{p2} + X_{m2} + \cdots + X_{pn} + X_{mn}, \tag{9.36}$$

where $X_{pi}$ and $X_{mi}$ are the paternal and maternal contributions to the genotypic value at the $i$th locus. Define

$$C_{ij} = \text{Cov}(X_{pi}, X_{pj}) = \text{Cov}(X_{mi}, X_{mj})$$
$$C'_{ij} = \text{Cov}(X_{pi}, X_{mj}) = \text{Cov}(X_{mi}, X_{pj}) \tag{9.37}$$

as the covariances in any generation before selection, and $C_{ij}^*$ and $C_{ij}'^*$ as the corresponding covariances after selection in any generation. Observe that

$$V_G(t) = 2 \sum_i C_{ii}(t) + 4 \sum_{i<j} C_{ij}(t)$$
$$= V_G \qquad + \quad \delta(t), \tag{9.38}$$

where $V_G(t)$ is the genetic variance before selection in generation $t$, $V_G$ is the genetic variance in linkage equilibrium, and $\delta(t)$ is the contribution to the variance due to linkage disequilibrium. (Note that $C'_{ii}(t) = 0$ since the population is in Hardy–Weinberg equilibrium in each generation before selection, and that any changes in $C_{ii}(t)$ due to changes in gene frequencies can be ignored if the number of loci is large enough.)

As a result of recombination

$$C_{ij}(t+1) = (1 - r_{ij})C^*_{ij}(t) + r_{ij} C'^*_{ij}(t), \tag{9.39}$$

where $r_{ij}$ is the recombination fraction between loci $i$ and $j$. Write $f_{ij}$ for the fresh covariance generated by selection in any generation, and assume that it is the same in coupling and repulsion so that

$$f_{ij}(t) = C^*_{ij}(t) - C_{ij}(t) = C'^*_{ij}(t). \tag{9.40}$$

(To justify this assumption we observe that $C_{ij}(t)$ remains very small, as will be shown in a moment.) Then

$$C_{ij}(t+1) = (1 - r_{ij})C_{ij}(t) + f_{ij}(t). \tag{9.41}$$

To derive $f_{ij}$ suppose that selection changes the phenotypic variance by an amount $\Delta(t)$ from $V(t)$ to $V^*(t) = V(t) + \Delta(t)$. If the phenotypic and genotypic values are approximately bivariate normal, so that the regression of $G$ on $Y$ is linear and homoscedastic, the change in the genetic variance is $h^4(t)\Delta(t)$. Thus

$$\delta^*(t) - \delta(t) = 8 \sum_{i<j} f_{ij}(t) = h^4(t)\,\Delta(t). \tag{9.42}$$

We now assume that all loci have equal effects, so that $f_{ij}(t)$ is the same for all pairs of loci. Thus

$$f_{ij}(t) = h^4(t)\,\Delta(t)/4n(n-1). \tag{9.43}$$

Substituting in eqn (9.41) we find that at equilibrium

$$\hat{C}_{ij} = \hat{h}^4\,\hat{\Delta}/4n(n-1)r_{ij}. \tag{9.44}$$

Adding over pairs of loci we find that

$$\hat{\delta} = 4 \sum_{i<j} \hat{C}_{ij} = \tfrac{1}{2}\hat{h}^4\,\hat{\Delta}/H, \tag{9.45}$$

where $H$ is the harmonic mean of the recombination fractions. If selection changes the phenotypic variance by a constant proportion, $V^*(t) = \beta V(t)$, then

$$\hat{\delta} = \tfrac{1}{2}\hat{h}^2(\beta - 1)\hat{V}_G/H. \tag{9.46}$$

This gives the quadratic in $\hat{V}_G$

$$\frac{(2H + 1 - \beta)}{2H}\,\hat{V}_G^2 + (V_E - V_G)\hat{V}_G - V_G V_E = 0 \tag{9.47}$$

which is the analogue of eqn (9.34).

It will be seen from eqn (9.44) that $C_{ij}$ is inversely proportional to the recombination fraction $r_{ij}$, and that the individual covariances, $C_{ij}$, are very small, being of order $1/n^2$, which means that the correlations are of order $1/n$; nevertheless $\delta$ is of order 1 since it is the sum of the covariances of all *pairs* of loci, whose number is of order $n^2$.

The main conclusion is that the effect of linkage can be summarized by $H$, the harmonic mean of the recombination fractions. Computer simulations with loci allocated at random on the linkage map (Bulmer 1974a) have shown that $H$ may be as low as 0.1 in *Drosophila*, which has only three large chromosomes with no crossing over in males, but that $H$ is likely to be about 0.4 (little below its value of 0.5 with no linkage) in the majority of species which have haploid numbers in double figures. Some typical values of $\hat{V}_G$ predicted from eqn (9.47) are shown in Table 9.5. The validity of these results has been confirmed by computer simulations (Bulmer 1974a, 1976b). The effect of linkage is unlikely to be important except in a few species, such as *Drosophila*, which have very few chromosomes.

TABLE 9.5

*Values of $\hat{V}_G$ predicted from eqn (9.47) when $V_G = V_E = 50$*

| $H$ | $\beta$  0.1 | 0.5 | 0.9 | 1.1 | 1.5 | 2.0 |
|-----|------|------|------|------|------|------|
| 0.1 | 21 | 27 | 41 | 71 | $\infty$ | $\infty$ |
| 0.4 | 34 | 39 | 47 | 53 | 82 | $\infty$ |
| 0.5 | 36 | 41 | 48 | 53 | 71 | $\infty$ |

For simplicity the above theory has been developed for loci without dominance, and from eqn (9.43) onwards for loci with equal effects. It seems likely, however, that it will remain at least approximately valid when these restrictions are removed, with the understanding that $V_G$ and $\hat{V}_G$ are replaced by $V_A$ and $\hat{V}_A$ respectively, and that $V_E$ is replaced by $(V_D + V_E)$.

## Epistasis

The main conclusion of the previous section is that any change in the mean of the offspring distribution resulting from selection of the parents is permanent since it is due to changes in gene frequencies, while changes in the variance and higher moments are temporary and disappear when selection is relaxed since they are due to linkage disequilibrium. We shall now show that if there are epistatic interactions between loci, part of the change in the mean may be due to linkage disequilibrium and will thus disappear when selection is relaxed. For further discussion see Griffing (1960).

We shall suppose that selection acts in a single generation, and we shall consider its effect on their children, grandchildren, and succeeding generations. The selection differential, $S$, is the change in the phenotypic mean among the parents due to selection, and it can be decomposed into components representing changes in mean values among the components of the phenotypic value:

$$S = S_A + S_D + S_{AA} + S_{AD} + S_{DD} + \cdots + S_E. \tag{9.48}$$

The response to selection among their children can be decomposed in a similar way:

$$R = R_A + R_D + R_{AA} + R_{AD} + R_{DD} + \cdots \tag{9.49}$$

To find the relationship between $R$ and $S$ we shall make the simplifying assumption that the regression of offspring on parent is linear for all components; no attempt will be made to justify this assumption, but I believe that it will serve as a good approximation to the true regression. Under this assumption we find from the covariances derived in Chapter 6 that the biparental regressions are

$$E(A_c|\text{parental values of } A, D, AA, \text{etc.}) = \tfrac{1}{2}(A_F + A_M)$$
$$E(AA_c|\text{parental values}) \qquad\qquad = \tfrac{1}{4}(AA_F + AA_M) \tag{9.50}$$
$$E(AAA_c|\text{parental values}) \qquad\qquad = \tfrac{1}{8}(AAA_F + AAA_M),$$

and so on; all regressions involving the child's dominance deviation or its interactions are zero. Since the selection differential is the change in the average mid-parental value, we find that

$$R_A = S_A$$
$$R_{AA} = \tfrac{1}{2}S_{AA} \tag{9.51}$$
$$R_{AAA} = \tfrac{1}{4}S_{AAA},$$

and so on, all other components of $R$ being zero.

We now consider the response to selection in the grandchildren. In the absence of linkage the regressions on grandparental values are

$$E(A_c|\text{grandparental values of } A, D, AA, \text{etc.}) = \tfrac{1}{4}(A_1 + A_2 + A_3 + A_4)$$
$$E(AA_c|\text{grandparental values}) \qquad\qquad = \tfrac{1}{16}(AA_1 + AA_2 + AA_3 + AA_4),$$

$$\tag{9.52}$$

and so on, where subscripts 1 to 4 refer to the four grandparents; all regressions involving dominance deviations are zero, but the regression coefficients for additive epistatic components are increased when there is linkage. If $R^*$ denotes the response to a selection differential $S$ in the grandparents, with no further selection among the parents, we find that

$$R_A^* = S_A$$

$$R_{AA}^* = \tfrac{1}{4}S_{AA} \tag{9.53}$$

$$R_{AAA}^* = \tfrac{1}{16}S_{AAA},$$

and so on, all other components of $R$ being zero. When there is linkage the coefficients, except the first one, are increased.

It can be concluded that only $S_A$, the selection differential in the additive component, leads to a permanent change in the mean. Selection differentials in additive epistatic components lead to a temporary change in the mean which is reversed when selection ceases, presumably because they are due to linkage disequilibrium. It is thus important to know whether epistatic components of variance contribute to the parent–offspring correlation.

As an example consider a simple model of epistatic interaction. Suppose that there are two loci, each with two alleles, with the following genotypic values:

|      | BB | Bb | bb |
|------|----|----|----|
| CC   | 1  | 0  | −1 |
| Cc   | 0  | 0  | 0  |
| cc   | −1 | 0  | 1  |

If the gene frequencies are $\tfrac{1}{2}$ at both loci, all the variance is additive × additive. However we select individuals, based on their genotypic values, we cannot change their gene frequencies, so that when selection is relaxed the population will eventually return to its original state. For example, if we select individuals with value 1, the gametes they produce will be $BC$ and $bc$ in equal numbers, and the genotypic values in the next generation are 0 and 1 with equal frequency. There has, however, been no change in the gene frequencies and so the population must return to its original state when selection is relaxed.

## Problems

1. 20 observations are made on a standard normal variate, and the average of the two largest values is calculated. Find the Expected value of this quantity to three decimal places (a) from eqn (9.11), (b) from eqn (9.11) with $p$ defined in eqn (9.12), (c) exactly, from tables of Expected values of order statistics.

2. A character is normally distributed with $m = V = 100$ and $h^2 = \tfrac{1}{2}$ in both sexes. 100 animals of each sex are measured, and the best five males and the best 50 females are chosen as parents. What is the Expected value of the character in the next generation?

3. Find recurrence relations for the mean and the additive genetic variance, analogous to eqns (9.26) and (9.28), when optimizing selection operates on one sex only.

4. Consider optimizing selection with the parameter values in Table 8.2. Find the equilibrium value $\hat{V}_A$ (a) in the absence of linkage, (b) when the harmonic mean recombination fraction is 0.25.

5. Derive the value of $\beta$ in a selection experiment in which the middle 20 per cent are selected in each generation. Find the equilibrium value of $\hat{V}_A$ when $V_A = V_E = 100$, $V_D = 0$, (a) in the absence of linkage, (b) when the harmonic mean recombination fraction is 0.1.

6. Find the relationship between $R^*_{AA}$ and $S_{AA}$ in eqn (9.53) when there is linkage.

# 10 Natural selection

*In the last chapter we considered the short-term effects of selection over a few generations, on the assumption that changes in gene frequencies were negligible. In this chapter we shall consider the long-term effects of selection in determining gene frequencies. We shall in particular be interested in the forces which determine the level of genetic variability in natural populations.*

## The effect of selection on gene frequencies

Let $w(y)$ be the fitness of an individual with phenotypic value $y$, and let $f(y)$ be the density function of phenotypic values before selection. To study the effect of selection on gene frequencies, consider the subset of individuals with a specified gene inherited from a specified parent at some locus, the residual genotype and the environmental contribution being left free to vary; in other words we are conditioning on the gene present at a particular position in the ordered genotype. Suppose that the mean value in this subpopulation (the conditional mean) is increased (or decreased) by an amount $a$ over the overall population mean; following the terminology of Fisher (1930) $a$ will be called the average excess of the gene. If the locus contributes only a small proportion of the total variance, then to a good approximation the distribution of $y$ in this subpopulation is simply shifted by $a$, and the density function in this subpopulation, before selection, is $f(y - a)$. The fitness of this gene, relative to a subpopulation with zero average excess, is therefore

$$W = \int w(y)f(y - a)\,\mathrm{d}y / \int w(y)f(y)\,\mathrm{d}y. \tag{10.1}$$

By expanding $f(y - a)$ in a Taylor series about $a = 0$, we find that, to order $a^2$,

$$W = 1 + Aa + \tfrac{1}{2}Ba^2, \tag{10.2}$$

where

$$A = - \int f'w\,\mathrm{d}y / \int fw\mathrm{d}y$$
$$B = \int f''w\,\mathrm{d}y / \int fw\,\mathrm{d}y. \tag{10.3}$$

The average fitness is $\overline{W} = 1 + \tfrac{1}{2}B\,\mathrm{Var}(a)$, since $E(a) = 0$. (Averaging is over different alleles at the same locus.) It is more convenient to work with a relative fitness which has an average value of unity, so we define

$$F \equiv W/\overline{W} = 1 + Aa - \tfrac{1}{2}B(a^2 - \mathrm{Var}(a)) + \mathrm{o}(a^2). \tag{10.4}$$

If $p$ is the frequency, before selection, of an allele with average excess $a$, the frequency of this gene after selection is

$$p^* = Fp. \qquad (10.5)$$

The change in the gene frequency as a result of selection is

$$\Delta p = p^* - p = (F - 1)p. \qquad (10.6)$$

If $f(y)$ is the normal density function with mean $M$ and variance $V$, and if $DM$ and $DV$ are the changes in the mean and variance as a result of selection, it is easily shown that $A$ and $B$ defined in eqn (10.3) are given by

$$A = DM/V$$
$$B = (DV + DM^2)/V^2. \qquad (10.7)$$

We shall now consider the average excess. Consider a gene inherited from the father at the $i$th locus. If the population is in Hardy–Weinberg and linkage equilibrium, the average excess of the gene is by definition equal to the main effect of that gene under the random mating decomposition defined in Chapter 4, so that $a = X_i$ in the notation of that chapter. If there is linkage disequilibrium, due to selection in previous generations, the average excess may also be affected by genes at other loci which are associated with it. We retain the random mating decomposition of the genotypic value and we suppose that there is no epistasis so that

$$Y = m + \sum_{i=1}^{2n} X_i + \sum_{i=1}^{n} X_{i,i+n} + E.$$

The mean square linear regression of $Y$ on $X_i$ is $m + \beta_i X_i$ where

$$\beta_i = \mathrm{Cov}(X_i, Y)/\mathrm{Var}(X_i) = \sum_{j=1}^{n} \mathrm{Cov}(X_i, X_j)/\mathrm{Var}(X_i). \qquad (10.8)$$

(Note that $\mathrm{Cov}(X_i, X_j) = 0$ for $j > n$ since the population is in Hardy–Weinberg equilibrium before selection, and that $\mathrm{Cov}(X_i, X_{j,j+n}) = 0$ since the distribution of dominance deviations before selection is unaffected by selection in previous generations.) If this can be taken as the true regression, the average excess is $a = \beta_i X_i$. In a model with the same allelic effects at all loci it follows from symmetry that the numerator and denominator of $\beta_i$ are the same at all loci, so that $\beta_i$ has the common value $\beta = \hat{V}_A/V_A$. In this symmetrical situation the additive effect must be multiplied by $\hat{V}_A/V_A$ to obtain the average excess.

### Directional selection

We shall now consider the effect of directional selection on gene frequencies. Under this form of selection $DM$ and $DV$ will be of the same order of magnitude, so that $Ba^2$ can be ignored by comparison with $Aa$ in eqn (10.2).

Suppose also that the population is in linkage equilibrium (so that we are considering only the effect of the first generation of selection) and that phenotypic values are normally distributed. Write $B_{ij}$ for the $j$th allele at the $i$th locus, $p_{ij}$ for its frequency and $\alpha_{ij} = X_i(B_{ij})$ for its additive effect. Then

$$\Delta p_{ij} = \frac{DM}{V}\, \alpha_{ij}\, p_{ij}. \tag{10.9}$$

Let us now consider the change in the mean due directly to these changes in gene frequencies; that is to say, we consider the change in the mean in a new population which is in Hardy–Weinberg and linkage equilibrium with gene frequencies changed from $p_{ij}$ to $p_{ij}^* = p_{ij} + \Delta p_{ij}$. Considering a two-locus model for simplicity we find that

$$\Delta M \equiv \sum Gp_{1i}^* p_{1j}^* p_{2k}^* p_{2l}^* - \sum Gp_{1i} p_{1j} p_{2k} p_{2l}$$

$$= \frac{DM}{V} \sum G(\alpha_{1i} + \alpha_{1j} + \alpha_{2k} + \alpha_{2l})p_{1i} p_{1j} p_{2k} p_{2l} + \text{smaller terms}$$

$$= \frac{DM}{V}\, E(GA) = \frac{DM}{V}\, V_A. \tag{10.10}$$

The above result can clearly be extended to any number of loci, and holds whether or not there is epistasis. But if we assume linearity of regression, we find from the parent–child covariance that the total change in the mean from one generation to the next as a result of selection is

$$\frac{DM}{V}\, (V_A + \tfrac{1}{2}V_{AA} + \tfrac{1}{4}V_{AAA} + \cdots). \tag{10.11}$$

Only the term in $V_A$ results from changes in gene frequencies; hence only this term represents a permanent change in the mean. The remaining terms involving epistatic variances result from linkage disequilibrium generated by selection and therefore disappear when selection is relaxed. This argument confirms by a rather different route the argument of the previous chapter.

The eventual effect of continued directional selection will be to drive all loci to fixation, though the process is too complex to study in detail. We shall now turn our attention to selection for an intermediate optimum value, and its converse, selection for extreme values. How would one expect gene frequencies to behave under these types of selection?

### Stability of equilibria under stabilizing and disruptive selection

In a wild population natural selection may take the form of selection for an optimal value of a character. If the mean phenotypic value is not at the optimum, there will be a short period of rapid directional selection until the two coincide; thereafter there will be selection for intermediate values and against extreme values without any change in the mean. This type of selection is called stabilizing selection; the interesting question is whether genetic variability can be maintained under these conditions. Alternatively, there might be selection in favour of extreme phenotypes, perhaps in two different niches, and against intermediate phenotypes; could this situation of disruptive selection persist indefinitely, or is it an unstable equilibrium?

We shall consider this problem under a very simple genetic model with $n$ loci with equal effects, without dominance or epistasis, and with two alleles at each locus. We may represent the genetic situation at the $i$th locus as follows:

| Genotype | $B_1 B_1$ | $B_1 B_2$ | $B_2 B_2$ |
|---|---|---|---|
| Frequency (before selection) | $p_i^2$ | $2p_i(1-p_i)$ | $(1-p_i)^2$ |
| Effect | $0$ | $c$ | $2c$ |

The additive effect of $B_1$ at the $i$th locus is $\alpha_{i1} = -c(1 - p_i)$, and $\text{Var}(\alpha_i) = c^2 p_i(1 - p_i)$. From eqn (10.4)

$$\Delta p_i = p_i(1 - p_i)[- A\beta c + \tfrac{1}{2}B\beta^2 c^2(1 - 2p_i)] \tag{10.12}$$

where $\beta = \hat{V}_A/V_A$ in a steady-state under continuous selection. At equilibrium $\Delta p_i = 0$, so that at all unfixed loci the equilibrium gene frequency $P$ satisfies

$$A - \tfrac{1}{2}B\beta c(1 - 2P) = 0. \tag{10.13}$$

Note that $A$ and $Bc$ are of the same order of magnitude so that both terms must be kept. Note also that $A$ need not be identically zero unless $P = \tfrac{1}{2}$; this is not at variance with the expectation that $DM = 0$ at equilibrium, since eqn (10.7) is only an approximation depending on normality. It can however be assumed with sufficient accuracy that

$$B = (DV + DM^2)/V^2 = DV/V^2. \tag{10.14}$$

We will now investigate the stability of the equilibrium with $p_i = P$ at all loci. To do this, suppose that it is subjected to a small perturbation so that $p_i = P + e_i$. In the next generation we may write the deviation from $P$ as $e_i' = e_i + \Delta p_i$, where $\Delta p_i$ is given by eqn (10.12). We must, however, take into account the fact that the perturbation may change the values of $A$ and $B$, defined by eqn (10.3), from their equilibrium values due to the change in the distribution of $y$. To a first approximation we may regard the distribution of $y$ as simply changed in mean by an amount $-2c \sum e_j$. From eqns (10.12) and (10.13) we find that, to order $c^2$,

$$e_i' = e_i + P(1 - P)\left[2c^2 \beta \frac{\mathrm{d}A}{\mathrm{d}M} \sum e_j - c^2 \beta^2 Be_i\right]. \tag{10.15}$$

(The change in $B$ can be ignored since its contribution is of order $c^3$.) If $f(y)$ is symmetrical then $\partial f/\partial M = -\partial f/\partial y$; assuming that the selection function $w(y)$

is unaffected by the change in the mean (frequency-independent or 'hard' selection), we find from the definitions of $A$ and $B$ in eqn (10.13) that

$$dA/dM = B - A^2 \simeq B. \tag{10.16}$$

Hence

$$e_i' = e_i + P(1 - P)c^2 \, \beta B[2 \sum e_j - \beta e_i]. \tag{10.17}$$

This system of equations is simplified by a linear transformation to a new set of variables

$$\bar{e} = \sum e_j/n = (\bar{p} - P)$$
$$d_i = (e_i - \bar{e}) = (p_i - \bar{p}). \tag{10.18}$$

The variable $\bar{e}$ is the deviation of the average gene frequency over all loci, $\bar{p}$, from the equilibrium value, $P$; $d_i$ is the deviation of the $i$th gene frequency from the average gene frequency. Writing $2nc^2 \, P(1 - P)\beta = \hat{V}_G$, the actual genetic variance at equilibrium, we find that

$$\bar{e}' = \left[1 + \hat{V}_G \, B\left(1 - \frac{\beta}{2n}\right)\right]\bar{e} \simeq (1 + \hat{V}_G B)\bar{e}$$
$$d_i' = (1 - \hat{V}_G \, \beta B/2n)d_i. \tag{10.19}$$

It can be concluded that if $B \simeq DV/V^2$ is negative, that is to say if selection tends to reduce the phenotypic variance (stabilizing selection), then $\bar{e}$ is stable but the $d_i$s are unstable; thus the mean will return to its equilibrium value after a perturbation but the individual gene frequencies will move away from $P$ towards the boundaries, some to fixation at 0 and others to fixation at 1, so that genetic variability will be lost. We may say that the equilibrium is stable in mean but unstable in variance. On the other hand, if $B \simeq DV/V^2$ is positive, that is to say if selection tends to increase the phenotypic variance (disruptive selection), the $d_i$s are stable but $\bar{e}$ is unstable; thus the gene frequencies will move away from $P$ in the same direction, which is determined by the initial perturbation. The subsequent behaviour of the system will require further analysis.

Thus an equilibrium which is stable in both mean and variance cannot exist under the above simple genetic model under any form of hard selection in which the fitness of an individual depends only on his own phenotypic value and is independent of the rest of the population. In many situations, however, there is some form of competition between members of the population, so that an individual's fitness may also depend on the total population size (density-dependent selection) and on the relative frequencies of the phenotypes of other individuals (frequency-dependent selection). Density dependence has little or no effect on the stability of genetic equilibria (though it is very important in stabilizing total population size), but frequency dependence may have a marked effect and is considered by some authors to be a major factor in the maintenance of genetic polymorphisms.

In the present context frequency-dependent selection would manifest itself through a change in the fitness function $w(y)$ as a result of the change in the mean due to the perturbation. Thus the change in $A$ due to the perturbation will have an additional contribution due to change in the fitness function, and eqn (10.16) should be replaced by

$$\frac{dA}{dM} = B - A^2 + C \simeq B + C \qquad (10.20)$$

where the additional term $C$ is defined as

$$C = \frac{1}{V} \int (y - M) \frac{\partial w}{\partial M} f dy / \int fw dy = \mathrm{Cov}\left(y, \frac{\partial w}{\partial M}\right) \bigg/ V\bar{w}. \quad 10.21)$$

Thus the equations (10.19) which determine stability are replaced by

$$\bar{e}' = [1 + \hat{V}_G(B + C - B\beta/2n)]\, \bar{e} \simeq [1 + \hat{V}_G(B + C)]\, \bar{e}$$

$$d_i' = (1 - \hat{V}_G\beta B/2n)d_i. \qquad (10.22)$$

The equilibrium is stable in variance, as before, when $B \simeq DV/V^2 > 0$ (disruptive selection), and it is stable in mean when $B + C < 0$. Thus it is stable in both mean and variance if $B > 0$ and $B + C < 0$, which of course requires that $C$ is negative; this can be interpreted as a requirement for sufficient frequency-dependent selection in favour of rare phenotypes (indicated by a negative value of $C$) to balance the tendency of disruptive selection (indicated by the positive value of $B$) to destabilize the mean.

*Simulation results*

To illustrate these results we shall consider again the computer simulations described on pages 156–7. In this simulation a character was assumed to be determined by 12 unlinked loci each with two alleles, + and −, contributing 1 and 0 respectively to the character; the phenotypic value was obtained by adding a random normal deviate with zero mean and variance 4. The frequency of the + allele in the initial population was taken to be $\frac{1}{4}$ at three loci, $\frac{1}{2}$ at six loci, and $\frac{3}{4}$ at the remaining three loci. In each generation there were 500 individuals of each sex before selection and 100 after selection; the selected parents were mated at random.

Several different selection procedures were used. In the *control* simulation 100 males and 100 females were chosen at random from the 500 individuals of each sex in the population. Two types of *stabilizing selection* were used. In the first type (*rank selection*) the 500 individuals of each sex were arranged in rank order by their phenotypic value and the middle 100 were selected; in the second (*value selection*) the 100 individuals of each sex with phenotypic values nearest 12 were selected. Similarly two types of *disruptive selection* were used. In the first (*rank selection*) the 500 individuals of each sex were arranged in rank order and the largest 50 and the smallest 50 were selected; in the second (*value selection*) the 100 individuals of each sex with phenotypic values furthest from 12 were selected.

Under all these selection regimes there is an equilibrium with $p_i = \frac{1}{2}$ at all loci. To predict the effect of selection on gene frequencies we must find the values of $B$ and $C$ which determine the stability of this equilibrium. If we assume normality they can be calculated explicitly, but it is sufficient for the time being to consider their sign. Under stabilizing selection it is clear that $B = DV/V^2$ is negative, while under disruptive selection $B$ is positive. Under selection by value a small change in the mean leaves the fitness function unchanged so that $C = 0$. Under selection by rank a small change in the mean shifts the fitness function by the same amount, so that $\partial w/\partial M = - \partial w/\partial y$; under normality and with truncation selection (in which $w$ is 0 or 1), it can be shown that $C = - B$, which implies neutral stability in mean.

TABLE 10.1

*Predicted stability and observed behaviour after 20 generations of selection*

| Type of selection | Stability in mean | Stability in variance | $\bar{p}$ | $\sigma^2(p)$ |
|---|---|---|---|---|
| Control | Neutral | Neutral | 0.528 | 0.043 |
| Stabilizing rank | Neutral | Unstable | 0.528 | 0.060 |
| Stabilizing value | Stable | Unstable | 0.499 | 0.057 |
| Disruptive rank | Neutral | Stable | 0.500 | 0.000 |
| Disruptive value | Unstable | Stable | 0.000 | 0.000 |

Table 10.1 shows the predicted behaviour of the system under different forms of selection, compared with its observed behaviour after 20 generations of selection. Stability in mean has been assessed by the average gene frequency, $\bar{p}$, and stability in variance by the variance of gene frequencies, $\sigma^2(p) = \sum (p_i - \bar{p})^2/12$. At the beginning of each simulation $\bar{p} = \frac{1}{2}$ and $\sigma^2(p) = 0.031$. In the control simulation random genetic drift has caused a small random movement of $\bar{p}$ and a small increase in $\sigma^2(p)$, as expected. The observed behaviour under selection in most cases follows the prediction from the stability analysis. The variance in the gene frequencies, $\sigma^2(p)$, has increased considerably more under both types of stabilizing selection than in the control simulation and has decreased to zero under both types of disruptive selection. The mean gene frequency, $\bar{p}$, has moved a small distance, apparently at random, under stabilizing rank selection, has remained very close to $\frac{1}{2}$ under stabilizing value selection and has been fixed at the boundary, $\bar{p} = 0$, under disruptive value selection. (In another simulation with disruptive value selection the gene frequencies were fixed at the other boundary, $\bar{p} = 1$.) There is, however, one exception. Under disruptive rank selection the mean gene frequency was

exactly $\frac{1}{2}$ after 20 generations and was clearly positively rather than neutrally stable. It was shown on pages 156–7 that linkage disequilibrium becomes so large under this type of selection that the system behaves as if it is controlled by a single locus with two alleles having values 0 and 12. The stability analysis outlined above is inappropriate since it assumes that the system behaves in a polygenic manner. It is easily verified that in a single locus situation in which equal numbers of both homozygotes are selected and mated at random there is a stable equilibrium with a gene frequency of $\frac{1}{2}$.

It is fairly easy to see intuitively the reason for stability or instability in mean. Under stabilizing value selection in which individuals with phenotypic values nearest 12 are selected, an external optimum is imposed on the system, whereas no preference is given to any particular value under stabilizing rank selection. Under disruptive value selection in which individuals with phenotypic values furthest from 12 are selected, the system will clearly be repelled from this value whereas no disadvantage is given to any particular value under disruptive rank selection. In fact disruptive value selection behaves, after the first few generations, like directional selection in which the direction is determined by random events during the first few generations.

## *Competition

It was mentioned earlier that competition between individuals is a possible cause of frequency-dependent selection leading to the maintenance of genetic variability. As a crude model of competititon we shall suppose that an individual with value $y$ has *absolute* fitness

$$w(y) = (a - bC_y) \exp - \tfrac{1}{2} \frac{(y - \theta)^2}{\gamma}. \tag{10.23}$$

The first term represents a linear decline in fitness with $C_y$, which is the number of effective competitors of an individual with value $y$. (To make biological sense we take $a > 1$, $b > 0$, and we suppose that $C_y < a/b$ so that $w > 0$.) The second term represents selection for the optimal value $\theta$. To be specific, we suppose that two individuals with values $y$ and $y'$ compete with each other to an extent $\exp - \tfrac{1}{2}(y - y')^2/\alpha$; if $N$ is the total population size and if $f(y)$, the phenotypic density function, is normal with mean $M$ and variance $V$, it follows that

$$C_y = N \int \exp - \tfrac{1}{2} \frac{(y - y')^2}{\alpha} f(y') \, dy' = N(\alpha/\alpha + V)^{1/2} \exp - \tfrac{1}{2} \frac{(y - M)^2}{(\alpha + V)}. \tag{10.24}$$

We shall now consider the equilibrium position with $M = \theta$ but with arbitrary variance $V$; the total population size at this equilibrium is determined by the condition that $\bar{w} = 1$. The quantities $B$ and $C$ which determine the stability of this equilibrium can be found by a fairly straightforward integration if $y$ is normally distributed. It will be found that $B + C$ is negative for all values of $V$, so that the equilibrium is stable in mean; this is not surprising since there is selection for an optimal value. The explicit expression for $B$ is

$$B = [r\gamma^{5/2}(\gamma + V)^{-3/2} - (\alpha + \gamma + V)](\alpha\gamma + \alpha V + 2\gamma V + V^2]^{-1}. \quad (10.25)$$

The behaviour of the system depends on the sign of $B$. When $\alpha/\gamma > (a - 1)$, which means that competition between individuals is weaker than optimizing selection, $B$ is negative for all positive values of $V$; hence all genetic variability will be eliminated. On the other hand, when $\alpha/\gamma < (a - 1)$, there is a unique positive root of the equation $B = 0$, say $V = V_0$, which satisfies

$$a^2\,\gamma^5 = (\gamma + V)^3\,(\alpha + \gamma + V)^2. \quad (10.26)$$

$B$ is negative when $V > V_0$ and $B$ is positive when $0 < V < V_0$. If $V_E > V_0$, genetic variability will be eliminated as before, since $B < 0$ for all attainable values of $V$. However, if $V_E < V_0$, the system will move to a position in which $V = V_0$ and $V_G = V_0 - V_E$. For if $V > V_0$ the system is unstable in variance and genetic variability will be lost until $V = V_0$; if $V < V_0$ then $V$ will increase if there is any additional genetic variability available since the equilibrium is still stable. Since $DV = 0$ when $B = 0$, the system evolves by shifting its variance until the direct effect of selection on the variance is minimized, in the same way that it evolves by shifting its mean until the direct effect of selection on the mean is zero.

We shall now consider briefly the extension of these ideas to competition between species. Consider two ecologically equivalent species with fitness function $w(y)$, which may take any form and may depend on the means and variances of the two species, provided only that it is the *same* function for both species. We also assume that there are $n$ additive genes with similar effects segregating for the character in each species, the frequencies of the $i$th gene in the two species being $p_{1i}$ and $p_{2i}$ respectively; since there is no hybridization it does not matter whether or not these genes are homologous. We now consider the nature of the equilibrium which exists when the two species are identical, that is to say with $p_{1i} = p_{2i} = P$ at all loci. We define $B$ and $C$ as before; because the two species are identical it is unnecessary to specify to which species these quantities refer. By analogy with eqn (10.18) we define $\bar{e}_1 = (\bar{p}_1 - P)$ and $d_{1i} = (p_{1i} - \bar{p}_1)$, and likewise for the second species. By the same argument as before, the recurrence relations for small perturbations analogous to eqn (10.22) are found to be

$$(\bar{e}_1 - \bar{e}_2)' = (1 + V_G B)(\bar{e}_1 - \bar{e}_2) \quad (10.27a)$$

$$(\bar{e}_1 + \bar{e}_2)' = (1 + V_G(B + 2C))(\bar{e}_1 + \bar{e}_2) \quad (10.27b)$$

$$d_{1i}' = (1 - V_G\beta B/2n)d_{1i} \quad (10.27c)$$

$$d_{2i}' = (1 - V_G\beta B/2n)d_{2i} \quad (10.27d)$$

From the first equation it follows that the species means tend to diverge if $B > 0$. This provides a genetic model for the phenomenon of character displacement which is sometimes observed when two closely related species have a partially overlapping range. When this happens, populations of the two species are very similar in the non-overlapping parts of the range where only one species is present, but they diverge and are easily distinguishable in the area of overlap where both species occur together. The implication of eqn (10.27a) is that character displacement is only likely to occur when the overall effect of selection is disruptive rather than stabilizing in nature.

It is realized that the models just considered are very crude representations of biological reality, but they have been discussed to indicate the way in which genetic and ecological models can be integrated in the study of evolutionary problems. For further discussion see Bulmer (1974b).

*General theory

So far in this section we have considered a model with only two alleles per locus. We shall now try to extend the theory and to show that similar conclusions can be drawn in a general situation with multiple alleles. To simplify the analysis we shall suppose that the fitness function $w(y)$ and the phenotypic density function are both symmetrical about zero, and that the latter is approximately normal with variance $V$. Under these conditions $A = DM = 0$ exactly, and $B = DV/V^2$ approximately.

Write $B_{ij}$ for the $j$th allele at the $i$th locus, and $p_{ij}$ for its frequency. We assume additive gene action so that the genotypic value can be written

$$G(B_{1j_1}, B_{2j_2}, \ldots B_{nj_n} / B_{1k_1}, B_{2k_2}, \ldots B_{nk_n}) = c_0 + \sum_i (c_{ij_i} + c_{ik_i}). \quad (10.28)$$

We call $c_{ij}$ the effect of the allele $B_{ij}$, and we write $m_i$ for the mean value of the effects $c_{ij}$ at the $i$th locus and $\mu_{ri}$ for the $r$th moment about the mean; thus

$$m_i = \sum_j c_{ij} p_{ij}$$
$$\mu_{ri} = \sum_j (c_{ij} - m_i)^r p_{ij}. \quad (10.29)$$

From eqns (10.4) and (10.6) we find that

$$\Delta p_{ij} = \tfrac{1}{2} B \beta_i^2 [(c_{ij} - m_i)^2 - \mu_{2i}] p_{ij} \quad (10.30)$$

where $\beta_i$, representing the effect of linkage disequilibrium, is given by eqn (10.8).

Computing the mean of the allelic effects with gene frequencies changed from $p_{ij}$ to $p_{ij} + \Delta p_{ij}$, we find that the change in the mean of the allelic effects at the $i$th locus is

$$\Delta m_i = \sum_j c_{ij} \Delta p_{ij} = \tfrac{1}{2} B \beta_i^2 \mu_{3i}. \quad (10.31)$$

The total change in the mean genotypic value is

$$\Delta M = 2 \sum_i \Delta m_i = B \sum_i \beta_i^2 \mu_{3i} \simeq B\beta^2 \sum_i \mu_{3i}. \tag{10.32}$$

These changes will be inherited permanently in future generations.

We have assumed that both the fitness function and the phenotypic density function are symmetrical about zero. We shall now interpret the above results in two situations which satisfy these conditions. We first suppose that $G$, the genotypic value is skew to the right (so that $\sum \mu_{3i} > 0$) and that $E$, the environmental deviation is correspondingly skew to the left in such a way that their sum $Y = G + E$, is symmetrical. In this case the mean genotypic value changes under selection, and the mean of the distribution is not in a steady state, despite the fact that there is no change in the mean phenotypic value. The reason for this behaviour is the non-linear regression of $G$ on $Y$ induced by the opposite skewness of $G$ and $E$, as discussed in Chapter 7.

In the second situation we suppose that $G$ and $E$ are both symmetrical, but that some of the loci are skew to the right and some to the left in such a way that $\sum \mu_{3i} = 0$. In this case there is no change in the overall mean under selection, but there are counter-balancing changes in the means at individual loci depending on their skewness. These changes will continue under continued selection until the skew loci either become symmetric or lose their variability.

The change in the variance of the allelic effects at the $i$th locus can be computed in the same way. It is found to be

$$\Delta \mu_{2i} = \sum_j (c_{ij} - m_i)^2 \Delta p_{ij} - (\Delta m_i)^2$$
$$= \tfrac{1}{2} B \beta_i^2 (\mu_{4i} - \mu_{2i}^2) - \tfrac{1}{2} B \beta_i^2 \mu_{3i})^2. \tag{10.33}$$

If we assume that all loci are symmetric so that the means do not change, then

$$\Delta \mu_{2i} = \tfrac{1}{2} B \beta_i^2 (\mu_{4i} - \mu_{2i}^2). \tag{10.34}$$

It is a standard result in statistical theory that $\mu_{4i} \geqslant \mu_{2i}^2$, equality only being possible for two alleles with equal frequencies. Thus when there are three or more alleles, stabilizing selection ($B < 0$) reduces and disruptive selection ($B > 0$) increases the genetic variance at each locus by acting directly on the allele frequencies. When there are only two alleles this cannot happen, since if the mean is kept fixed there are no degrees of freedom left to vary. In this case, as we have already seen, the loss or maintenance of genetic variability depends in the same way on the sign of $B$, but is brought about in a different way, by gene frequencies at different loci 'repelling' each other towards fixation at the boundaries or 'attracting' each other towards the centre in pairs in such a way as to keep the mean constant.

Finally we consider whether the system is stable in mean, that is to say whether the mean will tend to return to zero after a perturbation. We suppose that the genotypic value is symmetrically distributed, so that there is a steady state with the mean at zero. Let the gene frequencies be perturbed from $p_{ij}$ to $p_{ij} + e_{ij}$, thus causing the mean to change from zero to

$$M = 2 \sum_{i,j} c_{ij} e_{ij}. \tag{10.35}$$

This perturbation will cause $A$ to change from zero to

$$A = \frac{\mathrm{d}A}{\mathrm{d}M} \cdot M = (B + C)M.$$

In the next generation we find from eqns (10.4) and (10.6) that

$$e'_{ij} = e_{ij} + (B + C)M\beta_i(c_{ij} - m_i)(p_{ij} + e_{ij}) + \tfrac{1}{2}B\beta_i^2[(c_{ij} - m_i)^2 - \mu_{2i}](p_{ij} + e_{ij}). \tag{10.36}$$

Assume that the $c_{ij}$s and the $e_{ij}$s are small quantities of the same order of magnitude, $\varepsilon$, and ignore quantities smaller than $\varepsilon^2$, so that $(p_{ij} + e_{ij})$ can be replaced by $p_{ij}$. Multiplying both sides by $2c_{ij}$ and summing, remembering that $\Delta M$ defined in eqn (10.32) is zero and that

$$2(B + C) \sum_i \beta_i \mu_{2i} = V_G,$$

the genetic variance, we find that

$$M' = (1 + (B + C)V_G)M. \tag{10.37}$$

Thus stability in mean depends, as before, on the sign of $(B + C)$.


**Mutation-selection balance**

Stabilizing selection reduces genetic variability, but in natural populations this is countered in each generation by fresh variability generated by mutation. In this section we shall consider how much genetic variability is likely to be maintained by the balance between these two forces. In this context it seems unrealistic to assume that there are only two alleles at each locus, since modern understanding of the nature of the gene shows that each gene is composed of a large number of potentially variable nucleotide sites. We shall therefore consider a model proposed by Kimura (1965) to allow for many alleles at a single locus and extended by Lande (1976) to many loci. In this model there are an effectively infinite number of alleles at each locus.

Kimura considered a single locus without dominance, so that we may write

$$Y = X_p + X_m + E, \tag{10.38}$$

where $Y$ is the phenotypic value, $X_p$ and $X_m$ are the contributions from the paternal and maternal alleles, and $E$ is the environmental contribution. He supposed that mutation could produce an infinite sequence of alleles, and that when a mutation occurred it changed the value of the contribution of that allele by a small amount $\xi$ from $X$ to $X + \xi$, where $\xi$ is a random variable with density function $f(\xi)$, which is symmetrical about zero with variance $\sigma^2$. Kimura considered a continuous time model in which mutation occurs at a rate $u$ per generation (which is taken as the unit of time), and he supposed that the phenotypic value was subject to selection for an optimum value, taken to be zero, so that the fitness (Malthusian parameter) of an individual with value $y$ is $-\frac{1}{2}y^2/\gamma$. (This quadratic model in continuous time is equivalent to the usual model of optimizing selection with fitness $\exp -\frac{1}{2}y^2/\gamma$ in discrete time.)

We assume that $X_p$ and $X_m$ are independently and identically distributed, and we denote the density function of alleles with effect $x$ at time $t$ by $p(x, t)$. The rate of change of $p$ due to mutation is

$$\frac{\partial p}{\partial t} \text{ (due to mutation)} = -up + u \int_{-x}^{x} p(x - \xi, t) \, f(\xi) d\xi$$

(10.39)

$$= u\sigma^2 \frac{\partial p}{\partial x^2} + o(\xi^2),$$

by a Taylor series expansion about $\xi = 0$. The Malthusian fitness of an allele, say a paternal allele, with effect $X_p = x$, relative to an average allele is

$$E(\tfrac{1}{2}Y^2/\gamma | X_p = x) - E(-\tfrac{1}{2}Y^2/\gamma) = (4M^2 + V - (x + M)^2)/\gamma,$$

where $M$ and $V$ are the mean and variance of $p(x, t)$. Thus the rate of change of $p$ due to selection is

$$\frac{\partial p}{\partial t} \text{ (due to selection)} = \tfrac{1}{2}p(4M^2 + V - (x + M)^2)/\gamma.$$

(10.40)

In the steady state when $t \to \infty$, $M$ must be zero from symmetry. By combining eqns (10.39) and (10.40) we find the total change in $p$ due to mutation and selection, which must be zero in the steady state, so that

$$\frac{\partial p}{\partial t} = \tfrac{1}{2}u\sigma^2 \frac{\partial^2 p}{\partial x^2} - \tfrac{1}{2}p(x^2 - V)/\gamma = 0.$$

(10.41)

This equation is satisfied if $p$ is the density function of a normal distribution with zero mean and variance $V$, where

$$V = (u\sigma^2 \gamma)^{1/2}.$$

(10.42)

Thus in the steady state the balance between mutation and selection produces a normal distribution of allelic effects with zero mean and variance $V$; the genetic value, $X_p + X_m$, is normal with zero mean and variance $2V$ since the paternal and maternal contributions are assumed to be independent.

Kimura (1965) extended this model to many loci by ignoring linkage disequilibrium. We write

$$Y = \sum_i (X_{ip} + X_{im}) + E, \qquad (10.43)$$

where $X_{ip}$ and $X_{im}$ are the contributions from the paternal and maternal alleles at the $i$th locus. Write $u_i$ for the mutation rate and $\sigma_i^2$ for the variance of mutational changes at the $i$th locus, and suppose that fitness is $-\frac{1}{2}y^2/\gamma$ as before. If we assume that there is no linkage disequilibrium so that the allelic effects at different loci are independently distributed, the argument of the previous paragraph remains valid for each locus and we find that in the steady state the variance of allelic effects at the $i$th locus is given by eqn (10.32) with $u$ and $\sigma^2$ interpreted as $u_i$ and $\sigma_i^2$. The genetic variance is

$$V_G = 2 \sum_i \mathrm{Var}(X_{ip}) = 2 \sum_i (u_i \sigma_i^2 \gamma)^{1/2}. \qquad (10.44)$$

The above argument is only approximate since it ignores linkage disequilibrium; indeed even the single locus argument is only approximate since it ignores departures from Hardy–Weinberg equilibrium which are always present under selection in a continuous time model. To take departures from Hardy–Weinberg and linkage equilibrium into account we turn to the discrete time model considered by Lande (1976). Suppose that the phenotypic value can be decomposed in the same way as before (eqn (10.43)), and that in each generation events occur in the order selection, recombination, mutation, random mating leading to the next generation. Write

$$\begin{aligned} C_{ij} &= \mathrm{Cov}(X_{ip}, X_{jp}) = \mathrm{Cov}(X_{im}, X_{jm}) \\ \tilde{C}_{ij} &= \mathrm{Cov}(X_{ip}, X_{jm}) \end{aligned} \qquad (10.45)$$

for the variances and covariances before selection, and $C_{ij}^*$ and $\tilde{C}_{ij}^*$ for the corresponding quantities after selection; note that $\tilde{C}_{ij} = 0$ since there is always Hardy–Weinberg equilibrium before selection. We now assume that before selection the allelic values $(X_{1p}, X_{1m}, X_{2p}, \ldots, X_{nm})$ have a multinormal distribution; we suppose that $E$ is independently normal with variance $V_E$.

Consider now the effect of selection acting on the phenotypic value according to the fitness function $w(y) = \exp -\frac{1}{2}y^2/\gamma$. The phenotypic value will remain normal after selection with zero mean but with variance changed from $V$, before selection, to $V^* = V - V^2/(V + \gamma)$, after selection. The effect on the variances and covariances of the allelic values can be found from their regressions on the phenotypic value. The regression of $X_{ip}$ on $Y$ is

$$X_{ip} = \frac{R_i}{V} Y + e \qquad (10.46)$$

where

$$R_i = \sum_j C_{ij}$$

$$\mathrm{Var}(e) = C_{ii} - R_i^2/V. \tag{10.47}$$

The variance of $X_{ip}$ after selection is

$$C_{ii}^* = \frac{R_i^2}{V^2} V^* + \mathrm{Var}(e) = C_{ii} - R_i^2/(V + \gamma). \tag{10.48}$$

The covariances after selection can be found from the regressions of $(X_{ip} + X_{jp})$ or of $(X_{ip} + X_{jm})$ on $Y$. They are

$$C_{ij}^* = C_{ij} - R_i R_j/(V + \gamma)$$

$$\tilde{C}_{ij}^* = -R_i R_j/(V + \gamma). \tag{10.49}$$

Denoting values after selection and recombination by two stars, we find that

$$C_{ij}^{**} = (1 - r_{ij})C_{ij}^* + r_{ij} \tilde{C}_{ij}^* = (1 - r_{ij})C_{ij} - R_i R_j/(V + \gamma), \tag{10.50}$$

where $r_{ij}$ is the recombination fraction (with $r_{ii} = 0$). The effect of mutation is to add $u_i \sigma_i^2$ to $C_{ii}^{**}$. Hence we obtain the recurrence relationships for the variances and covariances in successive generations before selection:

$$C_{ii}(t + 1) = C_{ii}(t) - R_i^2(t)/[V(t) + \gamma] + u_i \sigma_i^2$$

$$C_{ij}(t + 1) = (1 - r_{ij})C_{ij}(t) - R_i(t)R_j(t)/[V(t) + \gamma], \quad i \neq j, \tag{10.51}$$

where

$$V(t) = V_G(t) + V_E = 2 \sum_i R_i(t) + V_E.$$

In the steady state we find that

$$-\hat{R}_i^2/(\hat{V} + \gamma) + u_i \sigma_i^2 = 0$$

$$r_{ij} \hat{C}_{ij} + \hat{R}_i \hat{R}_j/(\hat{V} + \gamma) = 0. \tag{10.52}$$

Hence

$$\hat{R}_i = [u_i \sigma_i^2 (\hat{V} + \gamma)]^{1/2}$$

$$\hat{C}_{ij} = -[u_i \sigma_i^2 u_j \sigma_j^2]^{1/2}/r_{ij}, \quad i \neq j.$$

To find the expressed genetic variance, $\hat{V}_G$, we observe that

$$\hat{V}_G = 2 \sum \hat{R}_i = 2\alpha(\hat{V}_G + V_E + \gamma)^{1/2} \tag{10.53}$$

where

$$\alpha = \sum (u_i \sigma_i^2)^{1/2}.$$

Hence

$$\hat{V}_G = 2\alpha(\alpha + (\alpha^2 + V_G + \gamma)^{1/2}. \qquad (10.54)$$

The above argument rests on the assumption of multivariate normality of the allelic effects. The validity of this assumption as an approximation is discussed by Lande (1976), Felsenstein (1977), and Fleming (1979). It seems likely to provide a good approximation provided that the correlations $C_{ii}/(C_{ii}C_{jj})^{1/2}$ are small. The approximate normality of the marginals is suggested by Kimura's single locus model. It will be seen from eqn (10.54) that the expressed genetic variability under selection–mutation balance is independent of the recombination fractions. The difference between Kimura's result (10.44), which ignores linkage disequilibrium, and Lande's result (10.54) is due to the difference between continuous and discrete time models; under Kimura's continuous time model the rate of change of the phenotypic variance due to selection is $dV/dt = -V^2/\gamma$, whereas under Lande's discrete time model the change in the variance in one generation is $DV = -V^2/(V + \gamma)$. Lande (1977) has also shown that eqn (10.54) remains valid under assortative mating.

To interpret the biological significance of these results consider a character determined by $n$ equally mutable loci, with $u_i = u$ and $\sigma_i^2 = \sigma^2$ for each locus. The rate of production of new genetic variability per generation by mutation is $\sigma_m^2 = 2nu\sigma^2$, so that $2\alpha^2 = n\sigma_m^2$. Thus $\alpha$ can be expressed in terms of the number of loci and of the experimentally measurable quantity $\sigma_m^2$. $V_G$ increases approximately as the square root of the number of loci. Lande (1976) reviews experimental evidence from which it seems likely that $n\sigma_m^2/V_E$ lies between 0.01 and 0.1 for a number of characters (though he now believes 0.01 to be a more likely value than 0.1 (personal communication)). Table 10.2 shows the expressed heritability predicted from eqn (10.54) under different intensities of selection with these values of $n\sigma_m^2/V_E$. It can be concluded that high heritabilities can be maintained by mutation even under rather strong stabilizing selection; thus mutation can be a powerful force for maintaining genetic variability in natural populations.

It is interesting to compare these results with the genetic variance which can

TABLE 10.2

*Expressed heritability predicted from eqn (10.54)*
*under mutation–selection balance*

| $n\sigma_m^2/V_E$ | $\gamma/V_E$ | |
| --- | --- | --- |
| | 1 (strong selection) | 10 (weak selection) |
| 0.01 | 0.17 | 0.32 |
| 0.1 | 0.43 | 0.61 |

be maintained by mutation–selection balance in a model with two alleles per locus. If we allow a mutation rate $u$ in both directions, eqn (10.12) becomes

$$\Delta p_i = p_i(1 - p_i)[-A\beta c + \tfrac{1}{2}B\beta^2 c^2(1 - 2p_i)] + u(1 - 2p_i). \qquad (10.55)$$

Suppose that the fitness function is $w(y) = \exp - \tfrac{1}{2}y^2/\gamma$ as before, and that the mean is zero when $p_i = \tfrac{1}{2}$ at all loci. At an equilibrium with $p_i = P_i$, the $P_i$s must be symmetrical about $\tfrac{1}{2}$, whence $A = 0$ exactly and $B = -1/(\hat{V} + \gamma)$ to an adequate approximation. Then

$$(1 - 2P_i)[-\tfrac{1}{2}\beta^2 c^2 P_i(1 - P_i)/(\hat{V} + \gamma) + u] = 0. \qquad (10.56)$$

Thus $P_i$ is either $\tfrac{1}{2}$ or it satisfies the quadratic equation

$$\beta^2 c^2 P_i(1 - P_i) = 2u(\hat{V} + \gamma). \qquad (10.57)$$

Suppose now that there is an even number of loci, and that eqn (10.57) holds at all loci, half of the gene frequencies being at the smaller and half at the larger root. Since $\beta = \hat{V}_G/V_G$, $V_G = 2nc^2 P_i(1 - P_i)$, we find that

$$\hat{V}_G^2 = 4nuV_G(\hat{V}_G + V_E + \gamma). \qquad (10.58)$$

If there is no linkage, we find from eqn (9.27) that

$$\hat{V}_G = 4nu(V_E + \gamma)/(1 - 8nu) \simeq 4nu(V_E + \gamma). \qquad (10.59)$$

If this equilibrium exists (with $0 < P_i(1 - P_i) < \tfrac{1}{4}$), then a stability analysis as in the last section shows it to be stable, and that any alternative equilibrium with $P_i = \tfrac{1}{2}$ at any locus is unstable. If $\gamma$ is so large (selection is so weak) that the equilibrium (10.57) does not exist, then the equilibrium with $P_i = \tfrac{1}{2}$ at all loci is stable. If there is an odd number of loci a similar conclusion holds, except that when eqn (10.57) exists the stable equilibrium is with eqn (10.57) at all loci but one, and with a gene frequency of $\tfrac{1}{2}$ at the remaining locus.

The main conclusion from this analysis is that much less variability is maintained by mutation–selection balance under this model than under the infinite alleles model. Suppose that $\gamma = 10V_E$ (weak selection) and that $4nu = 0.01$ (which requires 250 loci each with a mutation rate of $10^{-5}$); the expressed heritability predicted from eqn (10.59) under these rather favourable circumstances is only 0.10.

## *Geographic variation

We have so far considered the maintenance of genetic variability by intraspecific competition leading to frequency-dependent selection and by mutation. Another possible way of maintaining genetic variability is selection for different optimal values in different environments (or niches). In a single locus context this is known as multiple niche polymorphism. We shall now investigate a model with two niches, in each of which there is selection for an optimal value, the optimal value being $\theta_1$ in niche 1 and $\theta_2$ in niche 2; the fitness function in niche $i$ ($i = 1$ or 2) is $w(y) = \exp - \tfrac{1}{2}(y - \theta_i)^2/\gamma$. After selection we suppose that a fraction $m$ of each population migrates from one niche to the other; migration is followed by random mating within each niche.

Before selection in generation $t$ we assume that the phenotypic value in niche $i$ is approximately normally distributed with mean $M_i(t)$ and with variance $V(t) = V_A(t) + V_E$, the latter being the same in both niches; we are thus presupposing the normal theory of Chapter 8, the dominance variance being subsumed in $V_E$ if it is present. After selection, the mean and variance of the breeding values in niche $i$ are

$$M_{Ai}^*(t) = M_i(t) + (\theta_i - M_i(t))V_A(t)/(V(t) + \gamma)$$
$$V_A^*(t) = V_A(t) - V_A^2(t)/(V(t) + \gamma). \tag{10.60}$$

Migration involves mixtures of the distributions in the two niches in proportions $m:(1 - m)$; thus after selection and migration we find that the mean and variance of breeding values in niche 1 are

$$M_{A1}^{**}(t) = (1 - m)M_{A1}^*(t) + mM_{A2}^*(t)$$
$$V_A^{**}(t) = V_A^*(t) + m(1 - m)[M_{A1}^*(t) - M_{A2}^*(t)]^2. \tag{10.61}$$

In the next generation, in the absence of linkage,

$$M_i(t + 1) = M_{Ai}^{**}(t)$$
$$V_A(t + 1) = \tfrac{1}{2}V_A^{**}(t) + \tfrac{1}{2}V_A, \tag{10.62}$$

where $V_A$ is the additive genetic variance in the absence of selection. In the steady state we find that

$$(\hat{M}_1 - \hat{M}_2) = \frac{(1 - 2m)\,\hat{V}_A(\theta_1 - \theta_2)}{\hat{V}_A + 2m(V_E + \gamma)} \tag{10.63}$$

and that

$$(\hat{V}_A - V_A) + \frac{\hat{V}_A^2}{\hat{V}_A + V_E + \gamma} = \frac{m(1 - m)(\theta_1 - \theta_2)^2\,\hat{V}_A^2}{[\hat{V}_A + 2m(V_E + \gamma)]^2}. \tag{10.64}$$

Eqn (10.64) provides a way of determining $\hat{V}_A$ given $V_A$, the additive genetic variance in the absence of linkage equilibrium. In the above analysis $V_A$ has been taken as given. We must now enquire what level of genetic variability can be maintained under the balance of migration and selection. We put forward the hypothesis that genetic variability will, if possible, increase (or decrease) until there is no linkage disequilibrium; if on balance the joint effect of selection and migration is to decrease the phenotypic variance so that $\hat{V}_A < V_A$, then $V_A$ will decrease, whereas in the opposite case it will increase. Putting $V_A = \hat{V}_A$ in eqn (10.64), we find that

$$[V_A + 2m(V_E + \gamma)]^2 = m(1 - m)(\theta_1 - \theta_2)^2 (V_A + V_E + \gamma). \tag{10.65}$$

This equation has a unique positive real root when $(\theta_1 - \theta_2)^2 > 4m(V_E + \gamma)/(1 - m)$, and no positive root otherwise. We therefore suggest that $V_A$ will tend to the positive root of eqn (10.65) when it exists and will be eliminated otherwise. We shall now show that this in fact happens under both the two allele and the infinite allele model.

Consider first the two allele model of page 167; denote the gene frequencies in the two niches at the $i$th locus by $p_{1i}, p_{2i}$, and suppose that there are $n$ loci segregating. Suppose also for simplicity that the mean is $\frac{1}{2}(\theta_1 - \theta_2)$ when all gene frequencies are $\frac{1}{2}$. In the steady state we shall have $p_{1i} = \frac{1}{2} - \varepsilon, p_{2i} = \frac{1}{2} + \varepsilon$ at all loci. At this point

$$(M_1 - M_2) = 4nc\varepsilon$$
$$V_A = \frac{1}{2}nc^2 - 2nc^2 \varepsilon^2 \simeq \frac{1}{2}nc^2 \quad \text{if } \varepsilon \text{ is small.}$$

(10.66)

$\hat{V}_A$ and $\varepsilon$ can now be determined from eqns (10.63) and (10.64).

To see whether this equilibrium is stable, we first write down the recurrence relationships for the gene frequencies:

$$p'_{1i} = p_{1i} + m(p_{2i} - p_{1i}) + (1 - m)\Delta p_{1i} + m\Delta p_{2i}$$
$$p'_{2i} = p_{2i} + m(p_{1i} - p_{2i}) + m\Delta p_{1i} + (1 - m)\Delta p_{2i}$$

(10.67)

where $\Delta p_{1i}$ and $\Delta p_{2i}$ are the changes due to selection given by

$$\Delta p_{1i} = p_{1i}(1 - p_{1i})[-A_1\beta c + \frac{1}{2}B_1\beta^2 c^2(1 - 2p_{1i})],$$

(10.68)

with a similar expression for $\Delta p_{2i}$. We can evaluate

$$A_1 = DM_1/\hat{V} = m(\theta_1 - \theta_2)/[\hat{V}_A + 2m(V_E + \gamma)]$$
$$A_2 = DM_2/\hat{V} = -A_1$$
$$B_1 = B_2 = B = (DV + DM^2)/V^2 = A^2 - 1/(\hat{V} + \gamma).$$

(10.69)

We also note that

$$dA_1/dM_1 = dA_2/dM_2 = B - A^2 = -1/(\hat{V} + \gamma).$$

(10.70)

We now consider a small perturbation, $p_{1i} = \frac{1}{2} - \varepsilon + e_{1i}, p_{2i} = \frac{1}{2} + \varepsilon + e_{2i}$. As before we only retain linear terms in the perturbations, and coefficients of order $1/n$, remembering that $c$ and $\varepsilon$ are both of order $n^{-1/2}$. The results are more easily expressed in terms of $s_i = e_{1i} + e_{2i}$ and $d_i = e_{1i} - e_{2i}$. After some algebraic reduction we find the recurrence relations

$$\bar{s}' = (1 - \alpha - \hat{V}_A/(\hat{V} + \gamma))\bar{s}$$
$$\bar{d}' = (1 - 2m)(1 - \alpha - \hat{V}_A/(\hat{V} + \gamma))\bar{d}$$
$$(s_i - \bar{s})' = (1 - \alpha)(s_i - \bar{s})$$
$$(d_i - \bar{d})' = (1 - 2m)(1 - \alpha)(d_i - \bar{d})$$

(10.71)

where

$$\alpha = (\hat{V}_A - V_A)/2nV_A. \qquad (10.72)$$

Both $\bar{s}$ and $\bar{d}$ are stable since $\alpha$ is of order $1/n$. The critical condition for the stability of the whole system is that $(s_i - \bar{s})$ should be stable, or that $\alpha > 0$.

Thus the system is stable when $\hat{V}_A > V_A$. If $\hat{V}_A < V_A$ the system is unstable; in this case gene frequencies will tend to go to fixation, some at zero some at one, and so $V_A$ will decrease either until it is zero or until a situation is reached when $\hat{V}_A = V_A$. On the other hand, if $\hat{V}_A > V_A$, the system is stable, and there is room for new genetic variability to be generated by genes at the boundaries, if they exist; thus $V_A$ will increase until $\hat{V}_A = V_A$. The analysis has for simplicity assumed a symmetric situation in which the mean is $\frac{1}{2}(\theta_1 + \theta_2)$ when all the gene frequencies are $\frac{1}{2}$; this assumption can be dropped without altering the conclusion, though the analysis becomes more complex (Bulmer 1971).

We now turn to the infinite allele model of page 175. We suppose that events occur in the order selection, migration, recombination, mutation, random mating. Before selection in each generation suppose that the allelic effects have a multinormal distribution in each niche, with means $m_{1i}$ and $m_{2i}$ at the $i$th locus in niches 1 and 2, and with variances and covariances the same in both niches, denoted by the same symbols as before. The covariances after selection are given by eqn (10.49) as before. We write $s_i = (m_{1i} + m_{2i})$, $d_i = (m_{1i} - m_{2i})$. The quantity $\sum s_i = \frac{1}{2}(M_1 + M_2)$ is driven rapidly by selection to the equilibrium value $\frac{1}{2}(\theta_1 + \theta_2)$, but apart from this constraint the $s_i$s are free to vary at will without affecting the properties or the behaviour of the system. After allowing for selection, migration, recombination and mutation, we find the recurrence relationships for the $d_i$s and the variances and covariances:

$$d_i(t+1) = (1 - 2m)[d_i(t) + R_i(t)D(t)/(V(t) + \gamma)]$$

$$C_{ii}(t+1) = C_{ii}(t) - \frac{R_i^2(t)}{V(t) + \gamma} + m(1-m)$$

$$[d_i(t) + R_i(t)D(t)/(V(t) + \gamma)]^2 + u_i \sigma_i^2 \qquad (10.73)$$

$$C_{ij}(t+1) = (1 - r_{ij})C_{ij}(t) - \frac{R_i(t)R_j(t)}{V(t) + \gamma}$$

$$+ m(1-m)\left[d_i(t) + \frac{R_i(t)D(t)}{V(t) + \gamma}\right]$$

$$\times \left[d_j(t) + \frac{R_j(t)D(t)}{V(t) + \gamma}\right] \quad i \neq j,$$

where

$$D(t) = (\theta_1 - \theta_2) - (M_1(t) - M_2(t)) = (\theta_1 - \theta_2) - 2\sum d_i(t).$$

In the steady state we find that the following equations are satisfied:

$$\hat{d}_i = (1 - 2m)\hat{R}_i(\theta_1 - \theta_2)/[\hat{V}_G + 2m(V_E + \gamma)]$$

$$-\frac{\hat{R}_i^2}{\hat{V} + \gamma} + \frac{m(1 - m)\hat{R}_i^2(\theta_1 - \theta_2)^2}{[\hat{V}_G + 2m(V_E + \gamma)]^2} + u_i\sigma_i^2 = 0 \qquad (10.74)$$

$$r_{ij}\hat{C}_{ij} + \frac{\hat{R}_i\hat{R}_j}{\hat{V} + \gamma} - \frac{m(1 - m)\hat{R}_i\hat{R}_j(\theta_1 - \theta_2)^2}{[\hat{V}_G + 2m(V_E + \gamma)]^2} = 0, \quad i \neq j.$$

To find the variability which can be maintained by the balance between selection and migration in the absence of mutation we set $u_i = 0$. In this case the only solutions of the second and third equations in (10.74) are *either* with $\hat{C}_{ij} = 0$ for $i \neq j$ (no linkage disequilibrium) and with eqn (10.65) satisfied, *or* with $\hat{V}_G = 0$. Numerical evaluation of the recurrence relations (10.63) suggests that when a positive solution of eqn (10.65) exists, the system converges to this solution from any starting point.

Thus our conjecture seems to be generally valid that under selection–migration balance the genetic variability will tend to the positive root of eqn (10.65) when it exists and will be eliminated otherwise. This provides another plausible mechanism for maintaining genetic variability. For example, if $m = 0.05$, $\gamma = V_E$, $(\theta_1 - \theta_2)^2 = 10V_E$, the heritability predicted from eqn (10.65) is 0.50.

We have for simplicity considered a model with only two niches. Felsenstein (1977) has considered a biologically more interesting model for a continuous geographic cline. He postulates a one-dimensional geographical continuum of infinite length, along which the optimum value changes linearly so that $\theta(x) = \beta x$, where $x$ denotes position $(-\infty < x < \infty)$. Individuals migrate at random in each generation, the variance of the migration distance in the direction of the cline being $\sigma^2$. At equilibrium the mean value at any point must be equal to the optimal value, since edge effects have been eliminated by considering an infinite line. Felsenstein finds that the genetic variance maintained under the infinite allele model by selection–migration balance with no mutation is

$$V_A = \tfrac{1}{2}\beta^2 \sigma^2 + \tfrac{1}{2}\{\beta^4\sigma^4 + 4(V_E + \gamma)\beta^2\sigma^2\}^{1/2}. \qquad (10.75)$$

As before there is no linkage disequilibrium. A more detailed analysis of clinal models has been made by Slatkin (1978).

## Problems

1. Verify eqn (10.7).

2. Evaluate $B$ and $C$ for the four types of selection in Table 9.1. Evaluate $V_G$ and $\hat{V}_G$ for stabilizing value selection assuming the initial gene frequencies in the simulation (3 at $\tfrac{1}{4}$, 6 at $\tfrac{1}{2}$, and 3 at $\tfrac{3}{4}$). Hence find the predicted behaviour of the gene frequencies from eqn (10.22).

3. Consider the model for intraspecific competition in eqns (10.23) and (10.24). How much variability can be maintained under this model with $\alpha = 1$, $\gamma = 4$, and $a = 1.5$?

4. Consider a character subject to selection for an optimal value with $\gamma = 2V_E$. Find the expressed heritability, $\hat{V}_G/(\hat{V}_G + V_E)$, maintained by mutation (a) under the infinite alleles model with $n\sigma_m^2 = 0.05V_E$, (b) with two alleles per locus with $n = 100$ and $u = 10^{-5}$.

5. Consider selection for different optimal values in two niches with $m = 0.01$, $\gamma = 2V_E$, $(\theta_1 - \theta_2)^2 = 4V_E$. What is the heritability of the character at equilibrium?

# 11 Selection indices

*The objective in much animal and plant breeding work is to select the best individuals to breed from. In the simplest case individuals with the largest phenotypic values for some trait are selected, with the objective of increasing the mean value of that trait in future generations. But the economic value of a breed or variety will often depend on a number of different traits; how should one take them all into account in assessing candidates for selection? Even in assessing single traits, one may have several sources of information about the genetic values of the candidates; thus one may know the phenotypic values of their close relatives, which can be used to increase the accuracy of the prediction of the genotypic values of the individuals. How can this be done in an optimal way?*

*The problem of combining several sources of information in an optimal way in assessing candidates for selection has given rise to the theory of selection indices which will be discussed in this chapter. We shall begin by describing the theory in a general context before applying it to answer the questions posed above.*

### The general theory of selection indices

Many selection problems have the following form. $N$ candidates are available for selection of whom a fixed number $n$ are to be chosen. A number of scores, $\mathbf{Y} = (Y_1, Y_2, \ldots, Y_k)$, are known for each candidate. The objective is to choose candidates with a high value of some target variable, $W$, the 'true worth', which cannot be observed directly but about which partial information can be obtained from the observed scores $\mathbf{Y}$. For example, the individuals might be candidates for a fixed number of university places; the observed scores are marks in examinations already taken, the target variable is an overall assessment of merit at graduation, and the objective is to use the observed scores to choose candidates who are likely to do best at graduation. In the context of animal and plant breeding, the observed scores are phenotypic values, either on several traits in each candidate or on the same trait in several relatives of each candidate, the target variable is genetic worth, and the objective is to use the observed phenotypic values to choose candidates of high genetic worth as parents of the next generation.

We shall suppose that $\mathbf{Y}$ and $W$ have a known joint probability distribution with density function $f(\mathbf{y}, w)$. We can therefore compute the regression of $W$ on $\mathbf{Y}$:

$$E(W|\mathbf{Y}) = \int wf \, \mathrm{d}w / \int f \, \mathrm{d}w. \tag{11.1}$$

For a candidate with a specified value of **Y**, the regression function $\hat{W} = E(W|\mathbf{Y})$ is a sensible predictor of $W$. A possible selection rule is to compute the predictor $\hat{W}$ for each candidate, to arrange the candidates in rank order on this criterion, and to select the candidates with the largest values of $\hat{W}$. We shall now show that this selection rule is optimal in the sense that it maximizes the Expected value of $W$ in repeated sampling.

Suppose that $n$ individuals are chosen by some rule based only on their **Y** values. The conditional distribution of $W$ given **Y** is unchanged by knowing that the individual is in the selected group since the selection rule depends only on the **Y**s. Write $W_s$ and $\hat{W}_s$ for the average values of $W$ and $\hat{W}$ in the selected group. Then

$$E(W_s|\mathbf{Y} \text{ for the selected group}) = \hat{W}_s. \tag{11.2}$$

To find the Expected value of $W$ in the selected group in repeated sampling, we must take the Expected value of both sides over the distribution of the **Y**s in the selected group; averaging over samples we find that

$$E(W_s) = E(\hat{W}_s). \tag{11.3}$$

But $\hat{W}_s$ is maximized in any particular sample, and hence its average is maximized over all samples, by choosing the $n$ individuals with the highest values of $\hat{W}$.

A function of $Y$, such as $\hat{W}$, which is used to rank and select candidates is called a *selection index*. We have thus shown that the use of the regression predictor $\hat{W}$ as a selection index is the optimal selection rule if optimality is interpreted as maximization of the Expected value of $W$. This is usually the appropriate criterion of optimality in animal and plant breeding when the aim is to maximize average genetic progress.

In deriving optimality it was assumed that the same amount of information was available for each candidate, but the argument remains valid (despite the contrary assertion by Henderson (1973)) in the important case in which more information is available about some individuals than about others; it is still optimal to select candidates with the highest predicted values of $W$, using whatever information is available about each candidate.

It has also been assumed that a fixed number of candidates is to be selected. An alternative procedure considered by many authors (e.g. Cochran 1951) is to allow the number of candidates to vary from one occasion to the next subject to a fixed proportion, $p$, of candidates selected in the long run. Suppose that the selection procedure is repeated a large number of times, say $r$ times. In the whole group of $rN$ candidates we want to ensure that a fixed number $rNp$ have been selected in an optimal way; we also note that the frequency distribution in this large group will very nearly reproduce the underlying probability distribution. Thus the optimal procedure is to select those candidates whose regression predictor $\hat{W}$ exceeds some fixed value $c$ such that $Pr[\hat{W} \geqslant c] = p$. This procedure, which uses a fixed pass mark, will be more efficient than passing a fixed number of candidates on each occasion because it allows more candidates to be chosen in good than in bad years. However, it may be impracticable to allow the number of successful candidates to vary from year to year. Also the determination of the pass mark is much more sensitive to assumptions about the joint distribution of $Y$ and $W$ than is the calculation of the regression function $\hat{W}$. For these reasons a fixed success rate rather than a fixed pass mark is usually appropriate in breeding applications.

There is thus a strong argument for using the regression predictor $\hat{W} = E(W|Y)$ as a selection index. We shall now discuss the main properties of regression predictors. Write

$$W = \hat{W} + e, \tag{11.4}$$

where $e$ is the prediction error. Observe that $E(e|Y) = 0$ for all $Y$ from the definition of $\hat{W}$, so that $\hat{W}$ and $e$ are uncorrelated (though not necessarily independent). Hence

$$\text{Var}(\hat{W}) = \text{Var}(\hat{W}) + \text{Var}(e). \tag{11.5}$$

We also note that

$$E(W) = E(\hat{W})(= m, \text{say}). \tag{11.6}$$

To find the correlation between $\hat{W}$ and $W$, multiply both sides of eqn (11.4) by $\hat{W} - m$ and take Expected values; we find that

$$\text{Cov}(\hat{W}, W) = \text{Var}(\hat{W})$$
$$\rho(\hat{W}, W) = [\text{Var}(\hat{W})(\text{Var}(W)]^{1/2}. \tag{11.7}$$

We shall now show that no other predictor can have a higher correlation with $W$ than $\hat{W}$. Consider any other predictor, say $W^*$, multiply both sides of eqn (11.4) by $W^* - E(W^*)$ and take Expected values. Since $W^*$ depends only on $Y$ it is uncorrelated with $e$. Hence $\text{Cov}(W, W^*) = \text{Cov}(\hat{W}, W^*)$. Thus

$$\rho^2(W^*, W) = \frac{\text{Cov}^2(W^*, W)}{\text{Var}(W^*)\,\text{Var}(W)} = \frac{\text{Cov}^2(\hat{W}, W^*)}{\text{Var}(\hat{W})\,\text{Var}(W^*)} \frac{\text{Var}(\hat{W})}{\text{Var}(W)}$$

$$= \rho^2(\hat{W}, W^*)\,\rho^2(\hat{W}, W) \leqslant \rho^2(\hat{W}, W).$$

The equality can only be attained when $\rho^2(\hat{W}, W^*) = 1$, which implies that $W^*$ is a linear function of $\hat{W}$.

The correlation $\rho(\hat{W}, W)$ provides a useful measure of the efficiency of $\hat{W}$ in selecting for $W$. If we select candidates based on their values of $\hat{W}$, the Expected increase in $\hat{W}$ in the selected group is $I\sigma(\hat{W})$ where $I$ is the selection intensity defined in Chapter 8; from eqn (11.3) this is also the Expected increase in $W$ in the selected group. If the $W$ values were known and could be used directly as a basis for selection, the mean increase in $W$ in the selected group would be $I\sigma(W)$. Thus we may define the relative efficiency of $\hat{W}$ as a selection index compared with complete knowledge of $W$ as

$$\text{Relative efficiency} = \frac{I\sigma(\hat{W})}{I\sigma(W)} = \rho(\hat{W}, W). \tag{11.8}$$

(The selection intensity $I$ depends on the shape of the distribution from which values are selected as well as on the proportion selected; any differences in the values of $I$ in the numerator and denominator of eqn (11.8) due to a difference in the shape of the distributions of $\hat{W}$ and $W$ has been ignored.)

Consider now the special case in which the regression of $W$ on $\mathbf{Y}$ is linear. Multivariate normality is a sufficient, though not a necessary, condition for this to occur. By standard theory the explicit formula for the regression is

$$\hat{W} = m + \sum \beta_i [Y_i - E(Y_i)], \tag{11.9}$$

where

$$\beta = \mathbf{V}^{-1} \mathbf{c}$$

$$\mathbf{V} = \text{Var}(\mathbf{Y}) \tag{11.10}$$

$$\mathbf{c} = \text{Cov}(W, \mathbf{Y}).$$

The residual error variance is

$$\text{Var}(e) = \text{Var}(W) - \beta^T \mathbf{c} = \text{Var}(W) - \mathbf{c}^T \mathbf{V}^{-1} \mathbf{c}. \tag{11.11}$$

Hence

$$\text{Var}(\hat{W}) = \mathbf{c}^T \mathbf{V}^{-1} \mathbf{c}, \tag{11.12}$$

from which, together with eqn (11.7), $\rho(\hat{W}, W)$ can be calculated.

If the underlying distribution is multinormal we have the additional information that the prediction error, $e$, is distributed normally and independently of $\mathbf{Y}$ or $W$. In this case it is also useful to know that, if $L$ is any linear function of $\mathbf{Y}$, $\rho(L, W)$ measures the efficiency of $L$ as a selection index for $W$. Under multivariate normality the regression of $W$ on $L$ is linear:

$$E(W|L) = m + \frac{\text{Cov}(L, W)}{\text{Var}(L)} [L - E(L)]. \tag{11.13}$$

If we select individuals with large values of $L$ with selection intensity $I$, the Expected increase in $L$ in the selected group is $I\sigma(L)$, and so the Expected increase in $W$ is

$$\Delta E(W) = \frac{\text{Cov}(L, W)}{\text{Var}(L)} I\sigma(L) = \rho(L, W) I\sigma(W). \tag{11.14}$$

Thus

$$\text{Relative efficiency} = \frac{\Delta E(W) \text{ using } L}{\Delta E(W) \text{ using } W} = \frac{\rho(L, W)I\sigma(W)}{I\sigma(W)} = \rho(L, W). \quad (11.15)$$

Since both $L$ and $W$ are normal, the values of $I$ in the numerator and denominator of eqn (11.15) are the same.

If the regression of $W$ on $\mathbf{Y}$ is not linear we may define the linear predictor $\hat{W}_L$ as

$$\hat{W}_L = m + \sum \beta_i [Y_i - E(Y_i)]. \quad (11.16)$$

This linear predictor can be used as an approximation to the true regression; it is in fact the linear mean square regression, that is to say the linear predictor of $W$ which has the smallest mean square prediction error. If we use this predictor as a simple approximation, the results in eqns 11.5. 11.6, 11.7, 11.11, and 11.12 remain valid with $\hat{W}_L$ substituted for $\hat{W}$. (Note that $e$ and $\hat{W}_L$ are still uncorrelated although $E(e|\mathbf{Y})$ is no longer identically zero.) The correlation between this linear predictor and $W$ is maximal among the class of linear predictors, though not of course among all predictors (Rao 1965, p. 233). However, this correlation is no longer an exact measure of efficiency if the regression of $W$ on $\hat{W}_L$ is non-linear, and it is possible in theory that a better linear predictor could be found for predicting *large* values of $W$. In practice it seems likely that departures from linearity will be small and that the linear predictor (11.16) will be nearly optimal.

### Selection indices for multiple traits

The economic value of most agricultural species depends on a number of different traits, and it is desirable to take them all into account in selection. In this section we shall consider how selection indices can be used for this purpose under a number of different circumstances.

*Selection among pure lines*

As early as 1936 Fairfield Smith constructed an index for selecting varieties of wheat. Wheat is self-fertilizing so that different varieties can be regarded as pure lines which reproduce themselves and which are reproductively isolated from each other. We can thus write down a model in the form

$$Y_i = G_i + E_i,$$

where $Y_i$ is the phenotypic value of the $i$th trait, $G_i$ is the genotypic value (which depends on the variety) and $E_i$ is an environmental deviation (assumed to be independent of $G_i$). Suppose that the economic value of the wheat from a plot with phenotypic values $(Y_1, Y_2, \ldots, Y_k)$ for the $k$ traits is a linear function of the $Y_i$s:

$$\text{Economic value} = a_0 + \sum a_i Y_i,$$

where the $a_i$s are known constants. The economic value of a variety with genotypic values $(G_1, G_2, \ldots, G_k)$ is

$$W = a_0 + \sum a_i G_i.$$

The objective is to select varieties with a high value of $W$, which will give good economic value in the future, using observations on phenotypic values in the past. The appropriate selection index is the regression of $W$ on the phenotypic values. It will be assumed that the underlying distribution is multinormal so that the regression is linear; it is in any case unlikely to depart much from linearity. We will also ignore the problem of genotype–environment interaction, which may raise serious difficulties if variety A is better than variety B this year but worse next year.

The construction of an index presupposes that we are thinking of the varieties as being drawn at random from some distribution. We must first obtain estimates of the relevant variances and covariances from experimental data. As an example, Fairfield Smith considered a variety trial in which ten Australian varieties of wheat were grown in replicated square-yard plots, with six replicates for each variety, in 1933; there were 100 plants in each plot. Four traits were measured for each of the 60 plots:

$Y_1 = \log(\text{number of ears per plant})$
$Y_2 = \log(\text{average number of grains per ear})$
$Y_3 = \log(\text{average weight per grain})$
$Y_4 = \log(\text{weight of straw per plant}).$

Economic value is defined as $Y_1 + Y_2 + Y_3$ since this is the logarithm of the yield of grain per plant; the purpose of defining the $Y_i$s in logarithmic units was of course to ensure an additive relationship for value. The economic value of a variety is thus

$$W = G_1 + G_2 + G_3.$$

The multivariate analysis of variance on these data is shown in Table 11.1. The environmental variances and covariances, Var(E), can be estimated directly from the matrix of mean squares and products within varieties. The genetic variances and covariances, Var(G), can be estimated from the difference between the mean squares and products between and within varieties as shown at the bottom of the table.

Suppose now that we wish to construct an index to select between different varieties of wheat in a future experiment done under similar conditions and with $r$ replications for each variety, so that the $Y_i$s now represent mean values for $r$ plots. In the notation of eqn (11.10)

$$\mathbf{V} = \text{Var}(\mathbf{Y}) = \text{Var}(\mathbf{G}) + \frac{1}{r}\,\text{Var}(\mathbf{E})$$

$$\mathbf{c} = \text{Cov}(W, \mathbf{Y}) = \text{Cov}(W, \mathbf{G}) = \text{Var}(\mathbf{G})(1\ 1\ 1\ 0)^T. \cdot$$

Table 11.1

*Multivariate analysis of variance on four variables in ten varieties of wheat each replicated six times (Fairfield Smith 1936)*

| Source of variation | Degrees of freedom | Matrix of mean squares and products $\times 10^5$ | | | | Symbol | Expected value |
|---|---|---|---|---|---|---|---|
| Between varieties | 9 | 1987 | −1086 921 | −835 779 1252 | 231 130 293 515 | B | 6 Var(G) + Var(E) |
| Within varieties | 50 | 135 | −40 103 | 1 −16 17 | 80 27 −1 123 | W | Var(E) |
| Derived estimate of variance component | | 309 | −174 136 | −139 132 206 | 25 17 49 65 | (B−W)/6 | Var(G) |

(The vector of 1s and 0s at the end is in general the vector of $a_i$s in the definition of $W$.) The resulting selection indices and their efficiencies when $r = 1$ and 6 are as follows:

when $r = 1$
$$\hat{W} = \text{constant} + 0.59Y_1 + 0.42Y_2 + 1.08Y_3 - 0.23Y_4$$
$$\rho(\hat{W}, W) = 0.90$$

when r = 6
$$\hat{W} = \text{constant} + 0.84Y_1 + 0.67Y_2 + 1.12Y_3 - 0.12Y_4$$
$$\rho(\hat{W}, W) = 0.97.$$

It will be seen that $Y_3$ (weight per grain) receives greater weight than the other two components of yield because of its smaller environmental variability. The efficiency of the index without replication is 90 per cent so that there is little room for increasing efficiency by replication. If 60 plots are available in a future experiment it would probably be more efficient to compare 60 varieties of wheat without replication rather than 30 varieties replicated twice or 20 varieties replicated three times because the larger number of varieties to choose from would more than counterbalance the slight reduction in accuracy. (This argument presupposes that there is a very large pool of possible varieties to choose from.) It is also interesting to observe that the efficiency of the observed yield, $Y_1 + Y_2 + Y_3$, as a predictor of true yield is 82 per cent when there is no replication, compared with 90 per cent for the optimal index.

It should also be observed that $Y_4$ occurs in the selection index although it does not contribute to economic value, because it carries some additional information about $G_1 + G_2 + G_3$ even when $Y_1$, $Y_2$, and $Y_3$ are known; however, dropping $Y_4$ only has a small effect on the index (see Problem 1). Finally, we note that the constants in the indices cannot be evaluated without knowing the mean values of the variables. It is not necessary to know the constant to rank varieties on which the same number of observations have been made. However, it would be necessary to know these constants if we wanted to compare varieties with different numbers of replications, since the constants in the appropriate indices would be different.

## Selection in an outbred population

The above example relates to selection between pure lines, but the analysis can readily be extended to selection between individuals in an interbreeding population. This problem was first considered by Hazel (1943). We write

$$Y_i = A_i + E_i,$$

where $Y_i$ is the phenotypic value of the $i$th trait, $A_i$ is the corresponding breeding value (= additive genetic value), and $E_i$ includes both the environmental deviation and the dominance deviation; it will be assumed that there is no epistasis and that normal distribution theory can be applied. As before we assume that the economic value of an individual is a linear function of the $Y_i$s with known weights $a_i$. Since only breeding values are inherited from one generation to the next, we wish to select as parents the individuals with the highest values of

$$W = a_0 + \sum a_i A_i$$

since this will maximize the genetic gain in the next generation. To predict $W$ from the $Y_i$s we need to know the additive genetic variances and covariances, $\mathrm{Var}(\mathbf{A})$, as well as the phenotypic variances and covariances, $\mathbf{V} = \mathrm{Var}(\mathbf{Y})$. We can then compute

$$\mathbf{c} = \mathrm{Cov}(W, \mathbf{Y}) = \mathrm{Var}(\mathbf{A})\mathbf{a}^T, \tag{11.17}$$

where $\mathbf{a}^T = (a_1, a_2, \ldots, a_k)$. The regression of $W$ on $\mathbf{Y}$ can then be computed from eqns (11.9) and (11.10). As discussed in Chapter 6, the additive genetic variances and covariances can be estimated either from the parent–offspring regressions or from a multivariate analysis of variance on half-sib or full-sib families.

Hazel (1943) illustrated this procedure from data on pigs which will be re-analysed here. Three traits were measured:

$Y_1$ = weight at 180 days
$Y_2$ = a numerical score of market suitability assessed by a panel of judges
$Y_3$ = a measure of productivity, allowing for the size and weight of the litter and early mortality.

It seems reasonable to regard $Y_3$ as a character of the mother of the litter.

TABLE 11.2

*Estimates of phenotypic and genetic variances and covariances for three traits
in pigs (modified from Hazel 1943)*

| Source of variation | Symbol | | Weight | Score | Productivity |
|---|---|---|---|---|---|
| Phenotypic | Var(**Y**) | Weight | 1015 | 94 | † |
| | | Score | | 23 | † |
| | | Productivity | | | 94 |
| Genetic | Var(**A**) | Weight | 302 | 17 | 0 |
| | | Score | | 2 | 0 |
| | | Productivity | | | 15 |

† Not estimated.

Estimates of the phenotypic and genetic variances and covariances are given in
Table 11.2. The genetic variances and covariances have been estimated from
the regressions of offspring on dam as described in Chapter 6 for the first two
characters. No information was given by Hazel about the phenotypic
correlation between a pig's weight or score and her own productivity (as
opposed to that of her dam); the corresponding genetic correlations were taken
as zero since there was no evidence of their significance.

Analysis of the economic return from pig breeding had shown that the
relative economic values of the three traits were in the ratios $1:3:6$, so that the
target variable for prediction can be taken as

$$W = A_1 + 3A_2 + 6A_3.$$

An index is required for selecting young pigs for breeding purposes before their
own productivity has been measured; the information available about each pig
is its own weight and score, $Y_1$ and $Y_2$, and the productivity of its dam, denoted
by $Y_3^*$ to distinguish it from the pig's own productivity $Y_3$. The variance–
covariance matrix of $(Y_1, Y_2, Y_3^*)$ is obtained from Var(**Y**) in Table 12.2 by
substituting 0 for the asterisks. We can also compute

$$\mathbf{c} = \text{Cov}[W, (Y_1, Y_2, Y_3^*)] = (353\ 23\ 45),$$

after observing that $\text{Cov}(W, Y_3^*) = \frac{1}{2}\text{Cov}(W, Y_3)$. Computing the regression
from eqn (11.9) we find that the selection index is

$$\hat{W} = \text{constant} + 0.41Y_1 - 0.68Y_2 + 0.48Y_3^*.$$

The index based on weight and score alone is obtained by dropping the last
term because these variables are uncorrelated with the dam's productivity. The
efficiencies of indices based on different amounts of information are shown
below:

| Variables used in constructing index | $\rho(\hat{W}, W)$ |
|---|---|
| Weight, score, and dam's productivity | 0.40 |
| Weight and score | 0.37 |
| Weight | 0.36 |
| Score | 0.15 |
| Dam's productivity | 0.15 |

It will be seen that weight is much the most important and that the other two contribute little extra information.

### *Efficiency of index selection

Hazel and Lush (1942) compared the efficiency of index selection with that of two other methods, tandem selection and independent culling. In tandem selection only one trait is selected at any one time, but the chosen trait is varied from year to year; for example, we might select for trait 1 in year 1, trait 2 in year 2, and so on in a regular cycle. In independent culling a separate selection threshold is set for each trait, and animals are only selected if they are above the threshold for each trait simultaneously. Consider the simplest case of $k$ traits with the same economic value so that we may write

$$\text{Economic value} = Y_1 + Y_2 + \cdots + Y_k$$

$$\text{Genetic value} = W = A_1 + A_2 + \cdots + A_k.$$

Suppose also that the traits are independently and normally distributed with the same variance $V$ and the same heritability $h^2$. If we select a proportion $p$ from a large population, the selection intensity for index selection or tandem selection is $Z/p$, where $Z$ is the standard normal ordinate corresponding to $p$. Under independent culling we set the threshold for each trait in such a way that the probability of its being exceeded is $p^* = p^{1/k}$. This ensures that the proportion selected is $p$; the selection intensity for any trait considered by itself is $Z^*/p^*$, where $Z^*$ is the standard normal ordinate corresponding to $p^*$.

The genetic gain per generation is equal to the increase in $W$ as a result of selection. This can easily be calculated in the present simple case and is given below:

$$\text{Index selection:} \quad h^2 \frac{Z}{p} (kV)^{1/2}$$

$$\text{Tandem selection:} \quad h^2 \frac{Z}{p} V^{1/2} \tag{11.18}$$

$$\text{Independent culling:} \quad kh^2 \frac{Z^*}{p^*} V^{1/2}.$$

The efficiencies of the last two methods relative to index selection are:

$$\text{Tandem selection:} \quad k^{-1/2}$$

$$\text{Independent culling:} \quad k^{1/2}\, Z^*\, p/Zp^*.$$

(11.19)

For example, when $k = 4$ and $p = 0.1$, the efficiencies of tandem selection and independent culling relative to index selection are 0.50 and 0.80 respectively. For all $k$ and $p$ the genetic gain is in the rank order Index selection > Independent culling > Tandem selection. This conclusion can be extended to the more general case when the traits are correlated and of varying economic value and variability (Young 1961).

### *Non-linear selection indices

It is sometimes of interest to consider a criterion of economic value which is not linear in the traits measured. Thus an intermediate value of some trait may be optimal (for example, back-fat thickness in pigs to produce bacon which is neither too lean nor too fat), or the value of one trait may depend multiplicatively on the level of another trait. To develop the theory of selection indices in this situation we return to the case of selection between pure lines (for example, between varieties of wheat). It is probably sufficiently general to consider a quadratic value function, though the method to be discussed can be extended to any polynomial. We therefore define the worth of a line as

$$W = \sum a_i\, G_i + \sum b_{ij}\, G_i\, G_j.$$

The required selection index, $\hat{W}$, is the regression of $W$ on $\mathbf{Y}$. We suppose that $\mathbf{G}$ and $\mathbf{E}$ have independent multivariate normal distributions with known parameters. The regression of $G_i$ on $\mathbf{Y}$ is a linear function of $\mathbf{Y}$ which can be computed from standard regression theory and which will be denoted as

$$E(G_i|\mathbf{Y}) = L_i(\mathbf{Y}).$$

We may therefore write

$$G_i = L_i(\mathbf{Y}) + e_i$$

where $e_i$ is the deviation from the regression line; we know that the $e_i$s are multinormal with a dispersion matrix which can also be computed from standard theory, and we write

$$E(e_i\, e_j) = \sigma_{ij}.$$

We can now compute

$$E(G_i|\mathbf{Y}) = L_i$$
$$E(G_i\, G_j|\mathbf{Y}) = L_i\, L_j + \sigma_{ij}.$$

Hence

$$\hat{W} = \sum a_i L_i + \sum b_{ij} L_i L_j + \sum b_{ij} \sigma_{ij};$$

The third term is a constant independent of $\mathbf{Y}$; it can be ignored if the same information is available on all individuals.

Suppose for example that $Y_1$ and $Y_2$ are independent, standard normal variates with zero mean, unit variance and no phenotypic or genetic correlation. Let the heritabilities of the two characters be $h_1^2$ and $h_2^2$ respectively, and suppose that the worth of a line can be represented as $W = G_1 - G_2^2$. Then

$$\hat{W} = E(W|Y_1, Y_2) = h_1^2 Y_1 - h_2^4 Y_2^2 - h_2^2(1 - h_2^2).$$

The optimal selection procedure is to select the lines with the highest values of $\hat{W}$.

This non-linear index is optimal for selecting among pure lines, but care must be taken in extending the argument to an outbreeding population. In this case suppose that the economic value of an individual can be expressed as

$$\sum a_i Y_i + \sum b_{ij} Y_i Y_j,$$

and suppose that in the group of individuals selected as parents of the next generation the mean breeding value of the $i$th character is $\overline{A}_i$. (Strictly this is the average of the breeding values of sires and dams, $\overline{A}_i = \frac{1}{2}(\overline{A}_{iF} + \overline{A}_{iM})$.) The mean economic value of the population in succeeding generations as a result of selection in this generation is

$$\sum a_i \overline{A}_i + \sum b_{ij} \overline{A}_i \overline{A}_j + \sum b_{ij} \mathrm{Cov}(Y_i, Y_j).$$

Since selection does not cause a permanent change in the variances and covariances, genetic progress is maximized by maximizing

$$W = \sum a_i \overline{A}_i + \sum b_{ij} \overline{A}_i \overline{A}_j.$$

We now define the regression of breeding value on phenotypic value in the same way as before:

$$E(A_i|\mathbf{Y}) = L_i,$$

and we note as before that

$$E(A_i A_j|\mathbf{Y}) = L_i L_j + \sigma_{ij}.$$

If we select a group of $n$ individuals with mean phenotypic values $\overline{Y}_i$, then

$$\hat{W} = E(W|\overline{\mathbf{Y}}) = \sum a_i \overline{L}_i + \sum b_{ij} \overline{L}_i \overline{L}_j + \frac{1}{n} \sum b_{ij} \sigma_{ij}$$

To maximize genetic progress we want to choose the group of individuals with the highest value of $\hat{W}$.

Considering the same example as before, with economic value defined as $Y_1 - Y_2^2$, we want to find the group of $n$ individuals with largest value of $h_1^2 \overline{Y}_1 - h_2^4(\overline{Y}_2)^2$. This differs basically from the strategy for pure lines, which is to choose the individuals with the highest values of $h_1^2 Y_1 - h_2^4 Y_2^2$.

An alternative approach to characters with intermediate optima is to use a restricted selection index (Kempthorne and Nordskog 1959). Suppose that we have measured $k$ traits, and that we wish to improve the last $(k - r)$ traits while keeping the first $r$ traits at their present values which are satisfactory. Suppose that economic values can be assigned to the traits to be improved, and that the genetic merit of an individual can be expressed by the linear function

$$W = \sum_{r+1}^{k} a_i A_i.$$

We now try to find a linear index

$$L = \sum_{1}^{k} b_i Y_i$$

whose use will maximize $W$ but will on average leave $A_1, \ldots, A_r$ unchanged. Assuming normality we therefore seek $L$ to maximize $\rho(L, W)$ subject to the constraints that $\rho(L, A_i) = 0$ for $i = 1, \ldots, r$. Now

$$\rho(L, W) = \frac{\text{Cov}(L, W)}{\{\text{Var}(L)\text{Var}(W)\}^{1/2}} = \frac{\mathbf{a}^T \mathbf{V}^* \mathbf{b}}{\{\mathbf{b}^T \mathbf{V} \mathbf{b} . \mathbf{a}^T \mathbf{V}^* \mathbf{a}\}^{1/2}} \qquad (11.20)$$

where $\mathbf{V} = \text{Var}(\mathbf{Y})$ and $\mathbf{V}^* = \text{Var}(\mathbf{A})$. Since $\rho(L, W)$ is unaffected by multiplying the $b$s by a constant, we may take $\mathbf{b}^T \mathbf{V} \mathbf{b} = c$, a suitable constant. Thus the problem is to maximize $\mathbf{a}^T \mathbf{V}^* \mathbf{b}$ subject to the constraints

$$\mathbf{b}^T \mathbf{V} \mathbf{b} = c \qquad (11.21)$$

$$\mathbf{K} \mathbf{V}^* \mathbf{b} = 0 \qquad (11.22)$$

where $\mathbf{K}$ is an $r \times k$ matrix with $k_{ii} = 1$, $k_{ij} = 0$ for $i \neq j$. Eqn (11.22) expresses the constraints that $\text{Cov}(L, A_i) = 0$ for $i = 1, \ldots, r$. (Note that $\mathbf{a}$ is a vector of order $k$ with its first $r$ elements zero.)

We now introduce a Lagrange multiplier $\lambda$ corresponding to eqn (11.21) and a vector of $r$ Lagrange multipliers $\boldsymbol{\mu}$ corresponding to eqn (11.22). We want to maximize

$$\mathbf{a}^T \mathbf{V}^* \mathbf{b} - \lambda(\mathbf{b}^T \mathbf{V} \mathbf{b} - c) - \boldsymbol{\mu}^T \mathbf{K} \mathbf{V}^* \mathbf{b}.$$

Differentiating with respect to $\mathbf{b}$ we obtain the set of equations

$$\mathbf{a}^T \mathbf{V}^* - 2\lambda \mathbf{b}^T \mathbf{V} - \boldsymbol{\mu}^T \mathbf{K} \mathbf{V}^* = 0 \qquad (11.23)$$

which must be solved in conjunction with eqns (11.21) and (11.22). From eqn (11.23)

$$\mathbf{b} = \mathbf{V}^{-1} V^* (\mathbf{a} - \mathbf{K}^T \boldsymbol{\mu})/2\lambda \qquad (11.24)$$

Premultiplying by $\mathbf{K} \mathbf{V}^*$ and using eqn (11.22) we find an expression for $\boldsymbol{\mu}$; $\lambda$ can be chosen at will since it is a constant divisor of all the $b$s, and we choose $\lambda = \frac{1}{2}$. Substituting these values in eqn (11.24) we obtain the final result

$$\mathbf{b} = [\mathbf{I} - \mathbf{V}^{-1} \mathbf{V}^* \mathbf{K}^T (\mathbf{K} \mathbf{V}^* \mathbf{V}^{-1} V^* \mathbf{K}^T)^{-1} \mathbf{K} \mathbf{V}^*] \mathbf{V}^{-1} \mathbf{V}^* \mathbf{a}. \qquad (11.25)$$

*Sampling errors*

We shall finally consider briefly the effect of using estimates of genetic and phenotypic variances and covariances instead of their true values in constructing a selection index. This may be a serious problem in constructing indices for multiple traits because of the large number of parameters which must be estimated. If we assume multivariate normality and suppose that genetic worth is a linear function of breeding values, the optimal selection index will be a linear function of phenotypic values (11.9) which we denote by $L$. If we use this index the gain from unit selection intensity will be $\rho(L, W)\sigma_W$. If we use the estimated index $L^*$, the gain will be $\rho(L^*, W)\sigma_W$. The ratio $\rho(L^*, W)/\rho(L, W)$ is a measure of the relative efficiency of the estimated index $L^*$ compared with the optimal index $L$. In practice we must estimate the gain as $\rho^*(L^*, W)\sigma_W^*$ using the estimated rather than the true parameter values, and the deviation of this quantity from $\rho(L^*, W)\sigma_W$ is an error in predicting the genetic gain from selection.

Harris (1963, 1964) has presented the results of extensive computer simulations of an index with two traits of equal heritability and economic importance, some of which are shown below. The variances and covariances were estimated from a multivariate analysis of variance on data from simulated paternal half-sibships in which $s$ sires had each had $m$ progeny, all by different dams. Table 11.3 shows the relative efficiency of the estimated index, calculated as the average value of $\rho(L^*, W)/\rho(L, W)$. It will be seen that there may be a serious loss of efficiency unless sample sizes are rather large, which is most marked when heritability is low and when there is negative genetic correlation. (The environmental correlation is zero in Table 11.3, but other results show that positive environmental correlation has a detrimental effect on efficiency.) Harris also showed that there may be an appreciable error in predicting the gain from selection unless the sample size is rather large. It can be concluded that the construction of a useful index for combining multiple traits requires rather accurate measurements of genetic parameters obtained from a large sample of the order of 1000 individuals. This subject has been investigated more recently by Sales and Hill (1976).

If we are forced, through lack of information, to use estimates of the genetic

TABLE 11.3

*Relative efficiency of an estimated selection index*

| No. of sires | Offspring per sire | Heritability<br>Genetic correlation | 0.2<br>−0.5 | 0.2<br>0.5 | 0.5<br>−0.5 | 0.5<br>0.5 |
|---|---|---|---|---|---|---|
| 50 | 5 | | 0.55 | 0.71 | 0.83 | 0.94 |
| 50 | 20 | | 0.71 | 0.96 | 0.93 | 0.98 |
| 200 | 5 | | 0.77 | 0.96 | 0.96 | 0.99 |
| 200 | 20 | | 0.97 | 1.00 | 0.99 | 1.00 |

parameters with large sampling errors, it seems unlikely that the most efficient procedure is simply to substitute the estimates for the parameters in the index. It seems likely that some form of 'shrinkage' would be appropriate to allow for the errors in the estimates. This is a difficult statistical problem which requires further work. We shall meet this problem again at the end of the next section.

### Prediction of breeding values from multiple information

The second important application of selection index theory is in predicting the breeding value of a single trait from several different sources of information. We consider an outbreeding population and we write

$$Y = A + E$$

where $Y$ is the phenotypic value, $A$ is the breeding value, and $E$ includes both the environmental deviation and the dominance deviation; we assume that there is no epistasis and that normal distribution theory can be applied. Thus $A$ and $E$ are independently and normally distributed with means $m$ and $0$ and with variances $V_A$ and $V_E$ respectively. (Remember that $V_D$ is included in $V_E$; we denote $V_A + V_E$, the phenotypic variance, by $V$.) The environmental components in parents and offspring are independent, and the regression of the child's breeding value on those of his parents is

$$E(A_C | A_F, A_M) = \tfrac{1}{2}A_F + \tfrac{1}{2}A_M.$$

Since the environmental terms in parents and offspring are independent, the predicted response to selection is equal to the average selection differential among the breeding values of the parents:

$$E(Y_C - m) = E(A_C - m) = \tfrac{1}{2}E_S(A_F - m) + \tfrac{1}{2}E_S(A_M - m).$$

($E_S$ denotes an Expectation taken after selection has operated.)

To maximize the predicted response to selection we must maximize the selection differentials among the breeding values of the parents. These breeding values cannot be observed directly but must be estimated from phenotypic values. The more accurately the breeding values of potential parents can be estimated, the greater will be the selection differential among these breeding values for a fixed proportion of parents selected, and hence the greater the response to selection. An individual's breeding value can be predicted more accurately by using not only his own phenotypic value but also those of his relatives (parents, sibs, offspring) and also by making repeated measurements on characters such as milk yield of cows or fleece weight of sheep which can be measured more than once.

In this section we shall consider how this extra information can be used optimally by constructing a selection index and how much increase in accuracy it should provide. Suppose that $\mathbf{Y}$ is the vector of observed phenotypic values relevant to the prediction of the breeding value, $A$, of an individual and that we construct from them the selection index (regression predictor) $\hat{A}$ as described on page 186. The efficiency of this selection index based on the specified phenotypic values compared with perfect knowledge of the breeding value is $\rho(\hat{A}, A) = \sigma(\hat{A})/\sigma(A)$.

In the simplest situation selection is based only on the individual's own phenotypic value; this is usually called mass selection. In this case we find that

$$\hat{A} = m + h^2(Y - m)$$
$$\rho(\hat{A}, A) = h, \tag{11.26}$$

where $h^2 = V_A/V$ is the heritability. Selection based on $\hat{A}$ is equivalent to selection based on $Y$, so that it is unnecessary to compute $\hat{A}$ in this case. The relative efficiency is the square root of the heritability. When the heritability is high, mass selection is rather efficient and there is little room for improvement by taking relatives into account. The use of information from relatives is mainly of importance when the heritability is low and in selection for characters which can only be measured in relatives either because they are sex-limited (for example, the breeding value of a bull for use in a dairy herd can only be assessed from the milk yields of related cows) or because they can only be measured after death (carcass traits).

*Repeated measurements*

Suppose that $Y_1, Y_2, \ldots, Y_k$ represent repeated measurements on the same character in the same animal at successive times (for example, in successive years). As a simple model suppose that the environmental deviation consists of two components, a permanent component which contains the dominance deviation together with any environmental factors acting early in life and having a permanent effect on the animal, and a temporary component which represents short-term environmental effects and which changes randomly from one measurement to the next. If we call these components $P$ and $T$ respectively, we may write

$$Y = A + E = A + P + T,$$

and we may partition the environmental variance into corresponding components:

$$V_E = V_P + V_T.$$

We also assume that the breeding value is the same for all measurements on the same animal. The correlation between repeated measurements is called the repeatability, denoted $r$, and is given by the equation

$$r = (V_A + V_P)/V.$$

If we wish to use the measurements to predict the breeding value it is clear from symmetry that they should all receive equal weight, so that we can confine our attention to their mean $\overline{Y}$. By the standard technique for obtaining the variance of a mean we find that

$$\text{Var}(\overline{Y}) = V/b$$

$$\text{Cov}(\overline{Y}, A) = h^2 V$$

where

$$b = k/[1 + (k - 1)r].$$

From the results on page 189, the selection index for $A$ and its efficiency are

$$\hat{A} = m + bh^2(\overline{Y} - m)$$

$$\rho(A, \hat{A}) = hb^{1/2}. \tag{11.27}$$

On the other hand, if we wish to predict the permanent part of the phenotypic value, $Z = A + P$, as a guide to the future performance of the animal, we first note that $\text{Cov}(\overline{Y}, Z) = rV$. Hence the selection index and its efficiency are

$$\hat{Z} = m + rb(\overline{Y} - m)$$

$$\rho(Z, \hat{Z}) = (rb)^{1/2}. \tag{11.28}$$

Comparison of eqns (11.27) with (11.28) shows that animals will be ranked in the same order whether we are selecting for $A$ or $Z$.

As an example, consider the problem of selecting sheep for fleece weight, which has approximately a repeatability of 0.7 and a heritability of 0.4 (Turner and Young 1969). The increased accuracy as a result of selecting on two records rather than one can be found by comparing $b^{1/2}$ for $k = 2$ and 1, and is equal to $(2/1.7)^{1/2} = 1.08$. It would be uneconomic to wait until sheep are two years old before deciding whether to select them for breeding or culling; the small increase in accuracy from having two records would be more than counter-balanced by the increased generation length in a breeding programme (since the aim is to maximize the genetic gain per year rather than per generation) or by the extra time that unproductive sheep were left in the flock. However, multiple records will automatically become available for the older sheep, and it would be foolish not to take them into account by using the appropriate index.

Several problems may arise in applications. It will usually be necessary to correct the records for age and year effects, and possibly for other systematic effects, before using them. A more serious problem is that culling of young animals will bias the later records since they will not be a random sample from the original population (Curnow 1961a; Robertson 1966; Turner and Young 1969). It has also been assumed that records at different ages have the same variance and that the correlation between any two records is the same regardless of the time interval between them. Either of these assumptions may be wrong. For milk yield in cows, the yield in the first lactation is less variable and in consequence has a higher heritability than in subsequent lactations, probably because the condition of the cow is less variable at the first calving than at subsequent calvings (Johansson and Rendel 1968, p. 288). Furthermore the correlation between the yields in different lactations declines somewhat with the time between them. In this situation it is probably best to regard the milk yields in different lactations as different correlated traits and to construct an index by the method of the last section.

## Pedigree evaluation

The idea of assessing the worth of an animal for breeding purposes from the ancestors in his pedigree is well-known, but how the information from the different ancestors and the individual himself should be combined in an optimal way is less familiar, and breeders may be tempted to give too much weight to illustrious names in the remote ancestry and too little weight to the performance of the individual himself.

Suppose as an example that the phenotypic values of both parents and of the individual himself have been measured. It is clear from symmetry that the two parents should receive equal weight, so that $\hat{A}$ depends only on the individual's own phenotypic value, denoted by $Y_1$, and on the mid-parental value, denoted by $Y_2$. The variance–covariance matrix of $Y_1$ and $Y_2$ is

$$\mathbf{V} \equiv \begin{bmatrix} \mathrm{Var}(Y_1) & \mathrm{Cov}(Y_1, Y_2) \\ \mathrm{Cov}(Y_1, Y_2) & \mathrm{Var}(Y_2) \end{bmatrix} = \begin{bmatrix} 1 & \tfrac{1}{2}h^2 \\ \tfrac{1}{2}h^2 & \tfrac{1}{2} \end{bmatrix} V;$$

the covariances between $A$ and the $Y_i$s are

$$\mathbf{c}^T \equiv [\mathrm{Cov}(A, Y_1), \mathrm{Cov}(A, Y_2)] = (1 \quad \tfrac{1}{2})h^2 V.$$

From eqn (11.10) we find that $\hat{A}$ is given by the equation

$$\frac{(2 - h^4)}{h^2}(\hat{A} - m) = (2 - h^2)(Y_1 - m) + 2(1 - h^2)(Y_2 - m). \quad (11.29)$$

When $h^2$ is small $Y_1$ and $Y_2$ receive equal weight, but as $h^2$ increases $Y_1$ receives greater weight than $Y_2$. It may at first appear counter-intuitive that information on parents becomes less useful as heritability increases. The reason is that when heritability is high the individual's phenotypic value provides nearly perfect information about his breeding value; the parents' phenotypic values likewise provide nearly perfect information about their breeding values, but this only gives partial information about the breeding value of their offspring. When heritability is low, the error in using parental breeding values to estimate that of their offspring is negligible compared with the error in using phenotypic values to estimate breeding values.

The efficiency of the predictor is

$$\rho(A, \hat{A}) = h(3 - 2h^2)^{1/2}/(2 - h^4)^{1/2}. \quad (11.30)$$

The relative efficiency compared with mass selection is found by dividing this expression by $h$. The relative efficiency has its maximum value of 1.22 when $h^2$ is small, drops to 1.07 when $h^2 = \frac{1}{2}$ and to 1 when $h^2 = 1$. Thus knowledge of the phenotypic values of both parents adds less accuracy to the predictor than one might at first suppose.

This analysis can be extended to the general case in which any number of ancestors have been evaluated. In particular suppose that the phenotypic values of all direct ancestors for the last $k$ generations are known. Let $Y_{i+1}$ represent the average phenotypic value of the $2^i$ direct ancestors $i$ generations ago. $\hat{A}$ will depend only on the $Y_i$s ($i = 1, \ldots, k + 1$); note that $Y_1$ is the individual's own value. If we let $h^2$ tend to zero the matrix $V$ becomes diagonal, and it is easily shown that each of the $Y_i$s should receive equal weight and that the efficiency of the predictor is

$$\rho(A, \hat{A}) = h(2 - \tfrac{1}{2}^k)^{1/2}. \tag{11.31}$$

Thus even if we knew the phenotypic values of *all* an individual's direct ancestors, the relative efficiency compared with mass selection would only be 1.41, representing an increase in efficiency of 41 per cent; as the heritability increases the gain diminishes.

Thus if the individual's own phenotypic value is known, rather little extra information about his breeding value can be obtained from his ancestors. Parental values are most useful for sex-limited and carcass traits which cannot be measured in the individual himself. The efficiency of using the phenotypic value of one parent as a predictor of the breeding value is $\frac{1}{2}h$; this rises to $0.71h$ when the phenotypic values of both parents are known. If we use all the direct ancestors for the last $k$ generations, it can be shown as before that the efficiency for small $h^2$ is

$$\rho(A, \hat{A}) = h(1 - \tfrac{1}{2}^k)^{1/2}, \tag{11.32}$$

which tends to $h$ as $k$ becomes large. Thus if both parents are known, rather little extra information is provided by more remote ancestors even when the

TABLE 11.4

*Efficiency of prediction from pedigree evaluation*

| Information available | General formula | $h^2 \to 0$ | $h^2 = \frac{1}{2}$ | $h^2 = 1$ |
|---|---|---|---|---|
| Individual | $h$ | | $h$ | 0.71 | 1 |
| Individual + both parents | $h(3 - 2h^2)^{1/2}/(2 - h^4)^{1/2}$ | 1.22$h$ | 0.76 | 1 |
| Individual + all ancestors | | 1.41$h$ | | 1 |
| Individual + one parent | $h(5 - 2h^2)^{1/2}/(4 - h^4)^{1/2}$ | 1.12$h$ | 0.73 | 1 |
| Both parents | $h/2^{1/2}$ | 0.71$h$ | 0.50 | 0.71 |
| One parent | $h/2$ | 0.50$h$ | 0.35 | 0.50 |
| All ancestors | | $h$ | | 0.71 |

heritability is low; when the heritability is 1 the more remote ancestors provide no additional information. These results are summarized in Table 11.4.

### Sib and progeny testing

Information about ancestors, even when it is available, provides an estimate of an individual's breeding value of limited accuracy. More information can be obtained from observations on his sibs and his progeny. This is particularly valuable for carcass traits and for sex-limited characters which cannot be measured in the individual himself.

In pig breeding it is the practice in some countries to send two pigs from each litter to a testing station where they are reared under standard conditions and where carcass traits are measured after slaughter. These measurements are used as a basis for selection of the rest of the litter. In general, if $\overline{Y}$ is the mean phenotypic value of $n$ tested sibs and $A$ is the breeding value of an untested litter mate, then

$$\text{Var}(A) = h^2 V$$

$$\text{Cov}(A, \overline{Y}) = \tfrac{1}{2}h^2 V \qquad (11.33)$$

$$\text{Var}(\overline{Y}) = [1 + (n - 1)t]V/n$$

where $t$ is the correlation between sibs; in the absence of dominance and of effects due to common environment $t = \tfrac{1}{2}h^2$, but in the presence of these factors it is inflated. The regression of $A$ on $\overline{Y}$ is found in the usual way and is used as a selection index. When the number of animals tested varies in different litters this gives a logical way of deciding between litters about which different amounts of information are available. The efficiency is

$$\rho(\hat{A}, A) = \tfrac{1}{2}hn^{1/2}/(1 + (n - 1)t)^{1/2}. \qquad (11.34)$$

When $t = \tfrac{1}{2}h^2$ and when $n$ is large the efficiency has a maximal value of 0.71.

For many characters it is possible to measure all members of a litter without slaughtering any of them. Let $Y$ be an individual's own value and $\overline{Y}$ the mean value of his sibs. The best predictor of the breeding value is the joint regression of $A$ on $Y$ and $\overline{Y}$, which can be found from eqn (11.33) together with the obvious results

$$\text{Var}(Y) = V$$

$$\text{Cov}(Y, \overline{Y}) = tV$$

$$\text{Cov}(A, \overline{Y}) = h^2 V.$$

The method can be extended to more complicated situations in which, for example, both full and half sibs may have been measured.

Progeny testing means selecting parents on the basis of the performance of their offspring. It is particularly valuable in selecting for sex-limited characters, for example in selecting bulls in a diary herd based on the milk yields of their daughters. Suppose that a sire is mated to $n$ dams and has one daughter by each of them. Write $A$ for the breeding value of the sire, $Y_1$ for the average phenotypic value of the daughters (who are half sibs) and $Y_2$ for the average value of the dams. Then

$$\text{Cov}(A, Y_1) = \tfrac{1}{2}h^2 V$$

$$\text{Cov}(A, Y_2) = 0$$

$$\text{Var}(Y_1) = (4 + (n-1)h^2)V/4n$$

$$\text{Cov}(Y_1, Y_2) = \tfrac{1}{2}h^2 V/n$$

$$\text{Var}(Y_2) = V/n.$$

The regression of $A$ on $Y_1$ alone is

$$\hat{A} = m + b(Y_1 - m) \tag{11.35}$$

where

$$b = 2nh^2/(4 + (n-1)h^2). \tag{11.36}$$

The regression on $Y_1$ and $Y_2$, if information on dams is also available, is

$$\hat{A} = m + b^*(Y_1 - m) - \tfrac{1}{2}h^2 b^*(Y_2 - m) \tag{11.37}$$

where

$$b^* = 2nh^2/(4 + (n-1)h^2 - h^4). \tag{11.38}$$

The efficiencies of these predictors are $(\tfrac{1}{2}b)^{1/2}$ and $(\tfrac{1}{2}b^*)^{1/2}$ respectively. For example, when $n = 40$ and $h^2 = 0.25$, we find that $b = 1.4545$ and $b^* = 1.4612$, and that the efficiencies of the two predictors are 0.8528 and 0.8547 respectively; virtually no information is lost by ignoring the records of the dams. The efficiency tends to 1 for large $n$, so that the breeding value of the sire can be estimated as accurately as desired by measuring enough daughters by different dams.

## *Progeny testing of bulls

We shall finally consider in detail the progeny testing of bulls for milk yield in order to illustrate some of the complications which arise. Until about 1950 the standard way of assessing the result of a progeny test was to compute the daughter–dam comparison, the difference between the average value of the daughters of a particular bull and the average value of their dams:

$$\text{daughter–dam comparison} = (\text{daughter average} - \text{dam average})$$
$$= Y_1 - Y_2.$$

The rationale of this method is to obtain a more accurate estimate of the bull's worth than could be obtained from the daughter average by eliminating the influence of the dam; it also has the merit of eliminating differences between herds since the daughter is always raised in the same herd as her dam.

Unfortunately this method has three serious defects. Firstly, it over-corrects for the effect of the dam by giving daughter and dam equal weights, whereas from eqn (11.36) it will be seen that the dam should only receive a relative weight $\frac{1}{2}h^2$ compared with her daughter. If we ignore complications due to herd differences and assume random mating of bulls with cows, the correlation between the daughter–dam comparison and the bull's breeding value is

$$\rho = (nh^2)^{1/2}/[8 + (n - 5)h^2]^{1/2}.$$

With $n = 40$ and $h^2 = 0.25$, the efficiency of the daughter–dam comparison is 0.77, whereas an efficiency of 0.85 is obtained by using the daughter average alone. The second defect of the daughter–dam comparison is that it takes no account of the number of daughters tested, whereas less weight should be given to a progeny test based on a small number of daughters. Thirdly it takes no account of year differences between daughters and dams.

To overcome these difficulties the daughter–dam comparison was replaced in the 1950s by the contemporary comparison. In this method no attempt is made to correct for the dam's record, which contributes rather little information, but a correction is made for herd and year effects, and possibly for other fixed effects such as the age of the cow and the season of calving. The calculation is done in two stages: first correct the yields for herd–year (and possibly other) effects, and then find the regression of the bull's breeding value on the mean corrected yield of his daughters.

For the daughters of a particular bull in the $i$th herd–year, we calculate the difference

$$d_i = \text{daughter average} - \text{contemporary herdmate average},$$

where the contemporary herdmate average is the average of the daughters of other bulls in the same herd–year. The difference $d_i$ is an estimate of $\frac{1}{2}A$, where $A$ is the breeding value of the bull (not incorporating the mean). If there are $n_{1i}$ daughters and $n_{2i}$ contemporaries in the $i$th herd–year, the conditional variance of $d_i$ given $A$ is

$$\text{Var}(d_i|A) = \frac{V - \frac{1}{4}V_A}{n_{1i}} + \frac{V}{n_{2i}}. \tag{11.39}$$

If we define $w_i = 1/\text{Var}(d_i|A)$, the contemporary comparison is calculated as the weighted average

$$\text{Contemporary comparison} = C = \sum w_i d_i / \sum w_i. \tag{11.40}$$

The contemporary comparison is the minimum variance unbiased estimator of $\frac{1}{2}A$ among the class of estimators based on linear combinations of the $d_i$s. The conditional variance of $C$ is

$$\text{Var}(C|A) = 1/\sum w_i. \tag{11.41}$$

The unconditional variance of $C$ is

$$\text{Var}(C) = E\,\text{Var}(C|A) + \text{Var}\,E(C|A) = \frac{1}{\sum w_i} + \tfrac{1}{4}V_A. \qquad (11.42)$$

The second stage in the calculation is to find the regression of $A$ on $C$, which is

$$\hat{A} = bC \qquad (11.43)$$

where

$$b = \text{Cov}(A, C)/\text{Var}(C) = \tfrac{1}{2}V_A/\text{Var}(C). \qquad (11.44)$$

$\hat{A}$ is now used as an index for ranking and selecting bulls. Note that if the $n_{2i}$s are large compared with the $n_{1i}$s, the regression coefficient (11.44) is equivalent to eqn (11.36).

The contemporary comparison method as described above assumes that differences between herds and between years are entirely of environmental rather than genetic origin. It is probably true that differences in milk yield between herds are largely due to management; if we can estimate the proportion of the variability between herds that is of genetic origin (the between herd heritability), a correction can be made for this factor. Differences between years were probably largely environmental when the contemporary comparison was introduced, but increasing use of artificial insemination with intense selection pressure has hopefully led to an upward temporal trend in the genetic component of milk yield which should be allowed for in the analysis. Another defect of the method is that it fails to take into account the identities of the sires of the contemporaries. This was not relevant when it could be assumed that the contemporaries were all by natural service, so that their sires were a random sample from the herd, but it has become important with the increasing use of AI; a bull's contemporary comparison may be artificially decreased (or increased) if the contemporaries happen to have been sired by bulls who are better (or worse) than average. For these reasons the contemporary comparison has been replaced by the sire comparison method, which makes a direct comparison between the yields of the daughters of different sires and which allows for herd–year differences without making any assumption about their origin.

The statistical model can be written

$$y_{ij} = \alpha_i + \beta_j + e_{ij} \qquad (11.45)$$

where $y_{ij}$ is the average yield of the $n_{ij}$ daughters of the $j$th sire in the $i$th herd–year; $\alpha_i$ is thus a herd–year effect (also incorporating the mean), $\beta_j$ is half the breeding value of the $j$th bull tested, and $e_{ij}$ is an error term with variance

$$\text{Var}(e_{ij}) = (V - \tfrac{1}{4}V_A)/n_{ij}. \qquad (11.46)$$

The first step is to estimate the herd–year effects. To do this most efficiently we use the knowledge that the $\beta_j$s can be regarded as random variables with variance $\frac{1}{4}V_A$. We therefore re-write the model as

$$y_{ij} = \alpha_i + \varepsilon_{ij} \tag{11.47}$$

where $\varepsilon_{ij} = (\beta_j + e_{ij})$. Hence

$$\text{Var}(\varepsilon_{ij}) = \tfrac{1}{4}V_A + (V - \tfrac{1}{4}V_A)/n_{ij}$$

$$\text{Cov}(\varepsilon_{ij}, \varepsilon_{i'j}) = \tfrac{1}{4}V_A, \quad i \neq i' \tag{11.48}$$

$$\text{Cov}(\varepsilon_{ij}, \varepsilon_{i'j'}) = 0, \quad j \neq j'.$$

Least squares estimates $\hat{\alpha}_i$ can now be calculated by standard theory together with their variances and covariances.

We next compute $z_{ij} = y_{ij} - \hat{\alpha}_i$, which is an estimator of $\beta_j$; the variances and covariances of the $z_{ij}$s for fixed $\beta_j$ are

$$\text{Var}(z_{ij}|\beta_j) = \text{Var}(\hat{\alpha}_i) + (1 - 2c_{i,ij})\,\text{Var}(e_{ij}) \tag{11.49}$$

$$\text{Cov}(z_{ij}, z_{i'j}|\beta_j) = \text{Cov}(\hat{\alpha}_i, \hat{\alpha}_{i'}) - c_{i,i'j}\,\text{Var}(e_{i'j}) - c_{i',ij}\,\text{Var}(e_{ij})$$

where $c_{i,i'j}$ is the coefficient of $y_{i'j}$ in $\hat{\alpha}_i$. We now compute the minimum variance unbiased estimator of $\beta_j$, say $\hat{\beta}_j = \sum_i w_i z_{ij}$ (with $\sum_i w_{ij} = 1$) with variance $\text{Var}(\hat{\beta}_j|\beta_j)$. Finally we find the regression of $A_j$ on $\hat{\beta}_j$,

$$\hat{A}_j = b_j \,\hat{\beta}_j \tag{11.50}$$

where

$$b_j = \text{Cov}(A_j, \hat{\beta}_j)/\text{Var}(\hat{\beta}_j) = \tfrac{1}{2}V_A/[\text{Var}(\hat{\beta}_j|\beta_j) + \tfrac{1}{4}V_A]. \tag{11.51}$$

$\hat{A}_j$ is now used as an index for ranking and selecting bulls.

It must be mentioned that the method just described differs slightly from the best linear unbiased prediction (BLUP) advocated by Henderson (1963, 1973, 1977). The BLUP procedure estimates $\hat{\alpha}_i$ and $z_{ij}$ in the same way, but then ignores the sampling error in $\hat{\alpha}_i$, so that it takes

$$\text{Var}(z_{ij}|\beta_j) = \text{Var}(e_{ij}) \tag{11.52}$$

$$\text{Cov}(z_{ij}, z_{i'j})|\beta_j) = 0.$$

This procedure appears to me to lack logic, though it may make little difference in practice. It should also be mentioned that either of these methods presents computational problems if the number of observations is large.

## Optimal breeding strategies

The function of a selection index is to make optimal use of the information available about candidates for selection. In designing an animal or plant breeding programme it is also important to consider how many candidates to test, what proportion of them to select, and what information to collect about each of them. The economic cost of testing the candidates and of collecting the information must be assessed. If it is planned to test an individual's progeny or to take repeated measurements over several years, allowance must be made for the increased generation length before selection can occur. The objective is to maximize genetic gain per unit cost per year. The Expected value of this quantity can be calculated under different proposed breeding programmes, and a rational choice can be made between them on this basis. The details of the calculations will vary from case to case. In this section we shall discuss two situations in which rather general conclusions can be reached about the optimal allocation of resources.

### Selection of plant varieties

The optimal allocation of resources in plant breeding has been discussed by Finney (1958) and Curnow (1961b), who consider the following situation. A large number of new varieties has been developed by plant breeders. They are first subjected to preliminary trials in which unsuitable varieties are eliminated on the basis of resistance to pests and disease and similar characters. The remainder undergo more extensive trials in which yield is the main criterion for selection. Those with the highest yield are then compared with established varieties, and those found to be superior are passed into commercial use. The process may be schematically represented as divided into three phases:

Preliminary trials → Yield trials → Comparison with standard.

We shall only consider the second phase, the yield trials. We assume that the plants in each variety are genetically homogeneous: they may be pure lines, $F_1$ hybrids between pure lines, or vegetatively propagated cultivars.

Suppose that $N$ varieties enter the yield trials and that $n$ varieties are to be chosen from them, with the objective of maximizing Expected yield in the selected group. The trials may extend over several years, say $k$ years, with a fraction $P_r$ of the varieties tested in the $r$th year being selected for further testing in the next year ($r = 1, \ldots, k$). We may also choose only to test a randomly chosen fraction $P_0$ of the $N$ varieties in the first year, if the gain in accuracy among those tested more than counter-balances the reduction in the selection intensity. We are free to choose the proportions $P_r$ subject to the constraint that

$$P_0 P_1 \ldots P_k = n/N (= \pi, \text{ say}). \tag{11.53}$$

Suppose also that we may write the phenotypic value of a variety as

$$Y = G + E$$

where $G$ is normally distributed with mean $m$ and variance $V_G$ among the $N$ varieties before selection and $E$ is normally distributed with zero mean and variance $V_E$. The error variance is assumed to be inversely proportional to the area of land available for each variety, so that in the $r$th year of selection

$$V_E = P_0 P_1 \ldots P_{r-1} N\sigma^2/A_r \qquad (11.54)$$

where $A_r$ is the area of land used for plants in that year and $\sigma^2$ is the variance for unit area. In a continuous breeding programme there will, at any one time, be a cohort of varieties in their first year of testing, another in their second year, and so on, and the total area available, $A$, must be divided between these cohorts. We may therefore choose the areas $A_r$ subject to the constraint that

$$A_1 + A_2 + \cdots + A_k = A. \qquad (11.55)$$

The mathematical problem is to choose the $P_r$s and the $A_r$s in an optimal way subject to the constraints (11.53) and (11.55). The main weakness in the above formulation of the problem is that genotype–environment interactions are ignored. The normality of genotypic values is also questionable, but Curnow (1961) has shown that departures from normality make little difference to the answer.

We first consider selection in a single stage ($k = 1$). The only question in this case is whether we should test all the $N$ varieties available ($P_0 = 1$) or whether we should only test a random sample of them ($P_0 < 1$). By adopting the latter strategy we can reduce the error variance, but only at the expense of reducing the selection intensity. With arbitrary $P_0$, the predicted increase in the phenotypic value in the selected group is

$$\Delta E(Y) = IV_Y^{1/2}. \qquad (11.56)$$

$I$ is the selection intensity corresponding to the proportion selected $p = \pi/P_0$, which from eqn (9.11) can be approximated by

$$I = \phi(z)/p; \qquad (11.57)$$

the phenotypic variance is

$$V_Y = V_G + P_0 N\sigma^2/A. \qquad (11.58)$$

From normality the regression of $G$ on $Y$ is linear with slope $V_G/V_Y$, so that the predicted increase in the genotypic value is

$$\Delta E(G) = IV_G V_Y^{-1/2}. \qquad (11.59)$$

The derivative of eqn (11.59) with respect to $P_0$ is

$$\frac{d}{dP_0} \Delta E(G) = \frac{IV_G V_Y^{-3/2}}{P_0} \left\{ V_G + \tfrac{1}{2} \frac{P_0 N\sigma^2}{A} - \frac{z}{I} \left( V_G + P_0 \frac{N\sigma^2}{A} \right) \right\} \quad (11.60)$$

If eqn (11.60) evaluated at $P_0 = 1$ is positive, the optimal strategy is to test all $N$ varieties. Otherwise, one should test only a random sample, the proportion being calculated by equating eqn (11.60) to zero in order to maximize the

Expected gain; this gives an implicit equation in $P_0$ (which can be solved iteratively) since $z/I$ is also a function of $P_0$.

We shall now consider selection in two stages ($k = 2$). We assume that all varieties are tested in the first year ($P_0 = 1$), which is likely to be optimal since the selection intensity at each stage will be smaller than under single stage selection. We must therefore choose $P_1$ and $P_2$, subject to $P_1 P_2 = \pi$, and $A_1$ and $A_2$, subject to $A_1 + A_2 = A$, in such a way as to maximize the Expected genetic gain. By analogy with eqn (11.59), an approximate expression for the genetic gain is given by

$$E(G) = I_1 V_G V_{Y_1}^{1/2} + I_2 V_G^* V_{Y_2}^{1/2}. \tag{11.61}$$

$I_1$ and $I_2$ are the selection intensities corresponding to proportions selected $P_1$ and $P_2$, which can be approximated by eqn (11.57); $V_G^*$ is the variance of genotypic values after selection in the first year, which from Chapter 9 is given by

$$V_G^* = V_G \left\{ 1 - \frac{V_G}{V_{Y_1}} I_1(I_1 - z_1) \right\}; \tag{11.62}$$

the phenotypic variances are

$$V_{Y_1} = V_G + N\sigma^2/A_1$$
$$V_{Y_2} = V_G^* + P_1 N\sigma^2/A_2. \tag{11.63}$$

Numerical evaluation of eqn (11.61) has shown that the following simple rule is nearly optimal: make the proportions selected the same in both stages, and allocate the same amount of land to both stages, so that

$$P_1 = P_2 = \sqrt{\pi}$$
$$A_1 = A_2 = \tfrac{1}{2}A. \tag{11.64}$$

By analogy, it is likely to be nearly optimal in the general case with $k$ stages to take

$$P_1 = P_2 = \cdots = P_k = (\pi)^{1/k}$$
$$A_1 = A_2 = \cdots = A_k = A/k. \tag{11.65}$$

Two approximations have been made in evaluating the Expected genetic gain from eqn (11.61). Firstly, the selection intensity is only approximately given by eqn (11.67), even when the underlying distribution is normal, if a fixed number of varieties is selected rather than selecting varieties which exceed a predetermined yield; the error in this approximation is shown in Table 9.1. Secondly, the fact that the distribution of genetic values ceases to be normal after selection has been ignored, though the change in the variance due to selection has been taken into account. It is unlikely that either of these approximations will affect the conclusions significantly, but it is interesting to see how the second difficulty can be overcome by an elegant approach due to Cochran (1951).

Suppose that measurements $Y_1$ and $Y_2$ are made in the two stages of selection and that fixed yields, $c_1$ and $c_2$, are used as selection criteria, so that varieties with $Y_1 \geqslant c_1$ are chosen in the first stage and varieties with $Y_2 \geqslant c_2$ in the second stage. (This is equivalent to sampling with fixed proportions if the numbers of varieties selected in each stage are large.) Assume that in the absence of selection $G$, $Y_1$ and $Y_2$ would have a multinormal distribution and write $r_1, r_2$, and $r$ respectively for the correlation coefficients of $G$ with $Y_1$, of $G$ with $Y_2$, and of $Y_1$ with $Y_2$. To determine $c_1$ and $c_2$, let $Z_1$ and $Z_2$ be standard normal variates with correlation $r$, and find constants $k_1$ and $k_2$ such that

$$\text{Prob}[Z_1 \geqslant k_1] = P_1$$
$$\text{Prob}[Z_1 \geqslant k_1, Z_2 \geqslant k_2] = P_1 P_2. \tag{11.66}$$

$k_1$ can be found from tables of the normal distribution and $k_2$ can then be found from tables of the bivariate normal distribution. If $G$, $Y_1$, and $Y_2$ have zero means, then $c_1$ and $c_2$ are obtained by multiplying $k_1$ and $k_2$ respectively by the standard deviations of $Y_1$ and $Y_2$.

The regression of $G$ on $Y_1$ and $Y_2$ in the original population before selection may be written

$$G = b_1 Y_1 + b_2 Y_2 + e. \tag{11.67}$$

Taking the Expected value in the population after two stages of selection, and noting that $e$ is independent of $Y_1$ and $Y_2$, we have

$$E_s(G) = b_1 E_s(Y_1) + b_2 E_s(Y_2), \tag{11.68}$$

where $E_s$ denotes the Expected value after selection. To find the Expected values of $Y_1$ and $Y_2$ after selection, we first find the Expected value of $(Z_1 - rZ_2)$ after selection for $Z_1 \geqslant k_1$ and $Z_2 \geqslant k_2$. We find that

$$P_1 P_2 E_s(Z_1 - rZ_2) = \frac{1}{2\pi\sqrt{(1-r^2)}} \int_{k_2}^{\infty} \int_{k_1}^{\infty} (z_1 - rz_2) \exp -\tfrac{1}{2}(1-r^2)$$
$$\times (z_1^2 - 2rz_1 z_2 + z_2^2)\, dz_1\, dz_2$$
$$= \frac{\sqrt{(1-r^2)}}{2\pi} \int_{k_2}^{\infty} \exp -\tfrac{1}{2}(1-r^2)$$
$$\times (k_1^2 - 2rk_1 z_2 + z_2^2)\, dz_2$$

(since the integrand is an exact differential)

$$= (1-r^2)\,\phi(k_1)\, Q\left(\frac{k_2 - rk_1}{\sqrt{(1-r^2)}}\right) \tag{11.69}$$

(by completing the square and change of variable), where $\phi(.)$ is the standard normal density function and $Q(.)$ is the upper tail area of the normal distribution. The Expected value of $Z_2 - rZ_1$ can be found in the same way, and hence the Expected values of $Z_1$ and $Z_2$, and finally of $Y_1$ and $Y_2$ can be determined. Combining these results with eqn (11.68) and evaluating the regression coefficients, we obtain the final result

$$E_s(G) = \left[ r_1\,\phi(k_1)\,Q\left(\frac{k_2 - rk_1}{\sqrt{(1-r^2)}}\right) + r_2\,\phi(k_2)\,Q\left(\frac{k_1 - rk_2}{\sqrt{(1-r^2)}}\right) \right] \frac{V_G^{1/2}}{P_1 P_2}. \quad (11.70)$$

To apply this result in the present situation it remains to evaluate the correlations from the variances and covariances, which are:

$$\mathrm{Var}(G) = \mathrm{Cov}(G, Y_1) = \mathrm{Cov}(G, Y_2) = \mathrm{Cov}(Y_1, Y_2) = V_G$$

$$\mathrm{Var}(Y_1) = V_G + N\sigma^2/A_1 \quad\quad (11.71)$$

$$\mathrm{Var}(Y_2) = V_G + P_1 N\sigma^2/A_2.$$

In this formulation it has been assumed that selection at the second stage depends only on $Y_2$, the value observed at this stage, ignoring the yield in the first stage. It would be more efficient to base selection in the second stage on a weighted average of the two yields, $Y_2^* = w_1 Y_1 + w_2 Y_2$, with weights in the ratio of $P_1 A_1 : A_2$. The Expected gain under this procedure can be calculated as before from eqn (11.70), with the variances and covariances in eqn (11.71) replaced by

$$\mathrm{Var}(G) = \mathrm{Cov}(G, Y_1) = \mathrm{Cov}(G, Y_2^*) = V_G$$

$$\mathrm{Var}(Y_1) = V_G + N\sigma^2/A_1 \quad\quad (11.72)$$

$$\mathrm{Cov}(Y_1, Y_2^*) = \mathrm{Var}(Y_2^*) = V_G + P_1 N\sigma^2/(P_1 A_1 + A_2).$$

The theory of two-stage selection described above can be extended to any number of stages (multi-stage selection). It has been applied not only in selecting among plant varieties but also by animal breeders in outbred populations; for example, the animal breeder might want to select first for a character which can be measured cheaply, or early in life, and then to select for a character which is expensive to measure, or which can only be measured in adults. The theory can also be used to handle the problem presented by the culling of young animals discussed in the subsection on repeated measurements in the last section. For further discussion see Cunningham (1975).

*Optimal group size in progeny and sib testing*

In the discussion of single stage selection above we considered the optimal allocation of resources in selecting between genetically homogeneous plant varieties when a conflict arises between the number of varieties tested and the accuracy with which each variety can be assessed. On a fixed amount of land, is it better to test a large number of varieties with rather high sampling errors or a smaller number more accurately?

A similar problem arises in the design of animal breeding programmes. As discussed in the last chapter, the breeding values of candidates for selection can be assessed with greater accuracy from observations on their progeny and sibs, but there must be an upper limit to the total number of such observations which can be made. Is it better to test a small number of candidates very accurately, by obtaining information on large numbers of their progeny or sibs, or to test a larger number of candidates with rather limited information about each of them?

Robertson (1957) has discussed this question in the context of progeny tests of bulls in a dairy herd. Suppose that each test bull is mated to $n$ cows, and that the milk yield of one daughter from each mating is measured. Suppose that $S$ sires are to be selected for breeding purposes from the bulls tested, and that the total number of progeny measured is $N$. We shall regard $N$ and $S$ as fixed, because facilities only exist for testing $N$ progeny and because $S$ is the smallest number of sires required to avoid an unacceptable level of inbreeding. We are free to choose $n$, the number of progeny per bull, but choice of this number determines $N/n$, the number of bulls tested. How should we choose $n$ in order to maximize the Expected genetic gain?

If we use the optimal selection index the Expected gain from selection is

$$\Delta E(G) = I\rho(\hat{A}, A)\sigma(A). \tag{11.73}$$

The selection intensity $I$ can be approximated by $\phi(z)/p$, where $p$, the proportion selected, is given by

$$p = nS/N. \tag{11.74}$$

If we ignore the milk yields of the dams, so that the index is given by eqn (11.35), the square of the efficiency of the index is

$$\rho^2(\hat{A}, A) = nh^2/(4 + (n-1)h^2) = p/(p + c), \tag{11.75}$$

where

$$c = (4 - h^2)S/h^2\,N, \tag{11.76}$$

so that

$$I\rho(\hat{A}, A) = \phi(z)/[p(p + c)]^{1/2}. \tag{11.77}$$

Maximizing eqn (11.77) with respect to $p$ (using the fact that $d\phi/dp = z$), we find that the optimal value of $p$ satisfies the implicit equation

$$c = 2p(\phi - pz)/(2pz - \phi). \tag{11.78}$$

This equation can be solved iteratively to find $p$ given $c$, whence $n$ can be found.

From eqn (9.12) a better approximation to the selection intensity is given by $\phi(z^*)/p^*$, where

$$p^* = (2S + 1)p/2S. \tag{11.79}$$

If we follow through the same argument, we find that the optimal value of $p^*$ is given by eqn (11.78) if we replace $p$ by $p^*$, $z$ by $z^*$, and $c$ by $c^* = (2S + 1)c/2S$.

For example, Robertson (1957) relates that he was associated with a progeny-testing scheme in dairy cattle at an artificial insemination centre. In each generation two sires ($S = 2$) were to be selected on the basis of a total tested sample of 300 heifers ($N = 300$). The heritability of milk yield was assumed to be $h^2 = 0.25$. From eqn (11.76), $c = 0.1$; hence from eqn (11.78), $p = 0.092$, so that we should test 21 bulls on 14 daughters each. Using the more exact approach in eqn (11.79) we find that $c^* = 0.125$, $p^* = 0.103$, $p = 0.082$, so that we should test 25 bulls on 12 daughters each. The scheme actually chosen was to test 12 bulls on 25 daughters each, because of the cost of buying and keeping too many sires unemployed; to obtain a more precise analysis this extra economic factor should be introduced into the formulation of the problem.

The above analysis assumes that the selection of sires is based exclusively on the phenotypes of their offspring, and that these sires only have one offspring by each dam so that these offspring are all half sibs. These assumptions are particularly appropriate in the selection of bulls for milk yield since this character cannot be measured directly in the bulls, and since cattle normally have only one calf. Different selection schemes will be appropriate in different species, but a similar argument can often be used to determine the optimal group size. We shall consider one further example to illustrate the method.

In breeding pigs one might wish to select for a carcass trait which can only be measured after slaughter; sib-testing would be an appropriate method of selection. Suppose that a number of litters is available for testing, each sired by a different boar so that litter-mates are full sibs while members of different litters are unrelated. We shall assume that each litter consists of $m$ males and $m$ females, and that the objective is to select $S$ litters, the males from which will be used as boars in the breeding herds. We shall also assume for simplicity that the character can be tested equally well in females as in males, so that $n$ females are chosen for slaughter from each litter ($n \leqslant m$) without changing the number of males available for selection. We shall regard $S$, the number of litters to be selected, and $N$, the total number of pigs tested, as fixed. We are free to choose $n$, the number of pigs tested in each litter, but choice of this number determines $N/n$, the number of litters tested. How should we choose $n$ to maximize the Expected genetic gain?

The selection index to be used is the mean value of $n$ full sibs of the candidates for selection. The Expected genetic gain is given by eqn (11.73), with the proportion selected, $p$, defined as before, and with the efficiency, $\rho(\hat{A}, A)$, given by eqn (11.34). The simplest way of determining the optimal value of $n$ is to calculate $I\rho(\hat{A}, A)$ for different values of $n$ between 1 and $m$.

The formulation of the problem can be adapted to the situation in which the character can only be measured in males, so that if $n$ males in a litter are tested, only $(m - n)$ males are available for selection. In this case it seems sensible to require that the total number of individuals to be selected should be fixed at $T$, say, so that the number of litters to be selected is $T/(m - n)$. The Expected genetic gain is given by eqn (11.73), with $p(\hat{A}, A)$ given by eqn (11.34) but with the proportion selected given by

$$p = nT/(m - n)N. \qquad (11.80)$$

The value of $m$ which maximises the genetic gain can be found by trial and error.

## Problems

1. From Table 11.1 find the selection index for total yield and its efficiency based on the first three traits with $r = 1$.

2. From Table 11.2 find the selection index for the target variable given and its efficiency based on the score and the dam's productivity.

3. Fleece weight in sheep has a repeatability of 0.7 and a heritability of 0.4. One record is available on sheep A; it is one standard deviation above the mean. Three records are available on sheep B; they are 0.8, 0.9, and 1 standard deviation above the mean. Which sheep would you select for breeding purposes?

4. Find the formula for the selection index for breeding value if three records are available on the candidate and three records on its dam, with $r = 0.7$, $h^2 = 0.4$.

5. Find the formula for the selection index using the candidate's own value and the mean value of $n$ full sibs; assume that dominance and common environmental effects are absent.

6. A boar is mated to $n$ sows; from $m$ of the resulting litters two pigs are measured for some trait, while from the remainder only one pig is measured in each litter. Let $A =$ breeding value of the boar, $Y_1 =$ mean value for litters with two pigs measured, $Y_2 =$ mean value for other litters. Find the variances and covariances of these quantities, and hence write down the matrix equation to be solved to obtain the selection index for $A$; assume that dominance and common environmental effects are absent.

7. 100 varieties of wheat are available for trial in a field laid out in 100 plots. The genotypic values for yield are normal with unit variance, and the single plot error variance is 4. One variety is to be selected after the trial, which is to be completed in a single stage. Find the Expected genetic gain in yield if 100, 50, 25, 20, or 10 varieties are tested, with appropriate replication. What is the optimal number to test? (Use the exact values of $I$ in Table 9.1 or the approximation (9.12).)

8. In a progeny-testing scheme in dairy cattle, five bulls are to be selected for milk yield and facilities are available for recording the milk yields of 200 heifers. If the heritability of milk yield is 0.25, how many bulls should be tested?

# 12 Finite population size

*So far we have developed a deterministic theory which presupposes an effectively infinite population in which chance fluctuations can be ignored. In reality populations are always finite and may be small enough for chance fluctuations due to random sampling to be quite important. In this chapter we shall discuss some aspects of the stochastic theory of population genetics which are relevant to quantitative characters. We shall begin by considering the random fluctuations of gene frequencies in a finite population.*

## Random drift of gene frequencies

We assume a population with non-overlapping generations which maintains a constant size of $N$ diploid individuals. Consider a locus with two alleles, $B_1$ and $B_2$; denote the number of $B_1$ genes in the population in generation $t$ by $X_t$ and its relative frequency by $P_t$ $(=X_t/2N)$. We now imagine that each individual produces a large number of gametes, and that the next generation is produced by random union among these gametes. Thus given $P_t$, the gene frequency in generation $t$, the number of $B_1$ genes in the next generation, $X_{t+1}$, can be regarded as a binomial variate with probability $P_t$ and index $2N$, since each gene is chosen independently and at random from the effectively infinite pool of genes. This binomial sampling model was first considered explicitly by Wright (1931). It ignores the existence of two sexes and is thus most directly relevant to species without separate sexes; it also assumes that self-fertilization occurs with frequency $1/N$, whereas a more realistic model would prohibit selfing in an outbreeding species. Despite these shortcomings this simple model has proved extremely useful; it can often be related to more realistic but more complicated models through the concept of effective population size.

We shall now investigate the random fluctuations of gene frequencies under this model. To this end we write

$$P_{t+1} = P_t + dP \qquad (12.1)$$

where $dP$ is the change in the gene frequency in one generation of binomial sampling. Conditional on $P_t$, $dP$ is a random variable with zero mean, with variance $P_t(1 - P_t)/2N$, and with higher moments which are of order $1/N^2$ and which can therefore be ignored. We may therefore write to order $1/N$:

$$P_{t+1}^k = (P_t + dP)^k = P_t^k + kP_t^{k-1}\,dP + \tfrac{1}{2}k(k - 1)P_t^{k-2}\,(dP)^2. \qquad (12.2)$$

Taking Expected values conditional on $P_t$, we find that

$$E(P_{t+1}^k | P_t) = P_t^k + k(k-1)P_t^{k-1}(1-P_t)/4N. \qquad (12.3)$$

Taking Expected values in eqn (12.3) conditional only on the gene frequency in generation 0, we find that

$$E(P_{t+1}^k) = \left(1 - \frac{k(k-1)}{4N}\right)E(P_t^k) + \frac{k(k-1)}{4N}E(P_t^{k-1}). \qquad (12.4)$$

This series of difference equations can be solved successively for $k = 1, 2, 3,$ ..., with the initial values $E(P_0^k) = p^k$, where $p$ is the initial gene frequency. The first four moments are found to be

$$E(P_t) = p$$

$$E(P_t^2) = p - pq\left(1 - \frac{1}{2N}\right)^t$$

$$E(P_t^3) = p - \tfrac{3}{2}pq\left(1 - \frac{1}{2N}\right)^t + \tfrac{1}{2}pq(q-p)\left(1 - \frac{3}{2N}\right)^t$$

$$ \qquad (12.5)$$

$$E(P_t^4) = p - \tfrac{9}{5}pq\left(1 - \frac{1}{2N}\right)^t + pq(q-p)\left(1 - \frac{3}{2N}\right)^t$$

$$+ pq(pq - \tfrac{1}{5})\left(1 - \frac{3}{N}\right)^t,$$

where $q = (1-p)$. The first two moments are exact, and can be written in the form

$$E(P_t) = p$$

$$\qquad (12.6)$$

$$\mathrm{Var}(P_t) = pq\left[1 - \left(1 - \frac{1}{2N}\right)^t\right].$$

Exact values for the third and fourth moments are derived by Crow and Kimura (1970, §7.4).

These results are most easily interpreted if we imagine a very large population broken up into a large number of isolated groups (or lines) each containing $N$ individuals; there is random mating within each line and complete reproductive isolation between lines. Each line starts with the same initial gene frequency, $p$; eqn (12.6) shows the mean gene frequency (averaged over all lines) and the variance in gene frequency between lines after $t$ generations. The mean gene frequency remains constant because there is no active force (such as selection or mutation pressure) to move it consistently in one direction. The variance between lines increases with time because of the passive random movement of gene frequencies. Because frequencies of 0 and 1 represent absorbing barriers from which escape is impossible, we should expect that every line will ultimately become fixed at one or other of these values. This is confirmed by the fact that the limiting value of the variance as $t \to \infty$ is $pq$; a random variable which lies in the range 0 to 1 and which has mean $p$ and variance $p(1 - p)$ must take the values 0 and 1 with probabilities $(1 - p)$ and $p$ respectively.

Further light on the behaviour of this process can be obtained by considering the quantity

$$H_t = 2P_t(1 - P_t). \tag{12.7}$$

This quantity represents the frequency of heterozygotes in the population on the assumption that it is in perfect Hardy–Weinberg equilibrium; it thus seems a reasonable measure of the genetic variability within a line. (In fact, the Expected frequency of heterozygotes conditional on $P_t$ is $H_t \times 2N/(2N - 1)$ since two genes are sampled *without replacement* from $2N$ genes to form a diploid individual.) It is easily shown from eqn (12.5) that

$$E(H_t) = \left(1 - \frac{1}{2N}\right)^t H_0. \tag{12.8}$$

Thus genetic variability within a line declines, on average, at a rate of $1/2N$ per generation. It can also be shown from eqn (12.5) that

$$\text{Var}(H_t) = \tfrac{2}{3}H_0\left(1 - \frac{1}{2N}\right)^t + H_0(H_0 - \tfrac{2}{3})\left(1 - \frac{3}{N}\right)^t - H_0^2\left(1 - \frac{1}{2N}\right)^{2t} \tag{12.9}$$

These results can readily be generalized to a locus with several alleles. We write $P_i(t)$ for the frequency of the $i$th allele in generation $t$, and we suppose that the numbers of the different alleles in the next generation follow a multinomial distribution with these probabilities. We now write

$$P_i(t + 1) = P_i(t) + \mathrm{d}P_i, \tag{12.10}$$

where $\mathrm{d}P_i$ is a random variable with

$$E(\mathrm{d}P_i) = 0$$
$$\text{Var}(\mathrm{d}P_i) = P_i(t)[1 - P_i(t)]/2N \tag{12.11}$$
$$\text{Cov}(\mathrm{d}P_i, \mathrm{d}P_j) = -P_i(t)P_j(t)/2N, \quad i \neq j$$

and with negligible higher moments. The joint moments of the $P_i(t)$s can be found in the same way as before. The results (12.5) and 12.6) remain valid for any specified allele. If we define the heterozygosity as

$$H_t = \sum_{i \neq j} P_i(t)P_j(t) = 1 - \sum_i P_i^2(t), \tag{12.12}$$

it is easy to show that

$$E(H_{t+1}|H_t) = \left(1 - \frac{1}{2N}\right)H_t, \qquad (12.13)$$

so that eqn (12.8) remains valid. However, eqn (12.9) is no longer true. Note also that

$$E[P_i(t+1)P_j(t+1)|\mathbf{P}(t)] = P_i(t)P_j(t)\left(1 - \frac{1}{2N}\right), \quad i \neq j, \quad (12.14)$$

so that

$$\text{Cov}[P_i(t), P_j(t)] = p_i p_j \left[\left(1 - \frac{1}{2N}\right)^t - 1\right], \quad i \neq j. \qquad (12.15)$$

The problem can also be approached through calculating the inbreeding generated by random sampling from a finite population. Under Wright's binomial sampling scheme we may argue that two genes in generation $t$ have a chance $1/2N$ of being derived from the same gene in generation $t-1$ and a chance $1-1/2N$ of being derived from different genes; in the first case they are certainly identical by descent, and in the second case their chance of being identical is $f_{t-1}$, the inbreeding coefficient in generation $t-1$. (The chance of identity is the same for any pair of genes, whether they belong to the same or to different individuals.) Hence

$$f_t = \frac{1}{2N} + \left(1 - \frac{1}{2N}\right)f_{t-1}. \qquad (12.16)$$

With the initial condition that $f_0 = 0$, we find that

$$f_t = 1 - \left(1 - \frac{1}{2N}\right)^t. \qquad (12.17)$$

Identifying $(1 - f_t)$ with the heterozygosity, defined as the ratio of the Expected frequency of heterozygotes in generation $t$ to their Expected frequency in generation 0, we can reproduce eqn (12.8). However, this approach will not be pursued further since it seems less natural than the direct approach through the theory of stochastic processes. For a fuller account of this and other approaches to genetic drift the reader is referred to Crow and Kimura (1970).

We shall now consider a model which allows for separate sexes. We assume that the population maintains a constant size of $M$ males and $F$ females in each generation. Consider a locus with two alleles, $B_1$ and $B_2$, and denote the numbers of $B_1$ genes among males and females in generation $t$ by $X_t^*$ and $X_t^{**}$ respectively, and their relative frequencies by $P_t^*$ $(=X_t^*/2M)$ and $P_t^{**}$ $(=X_t^{**}/2F)$. We imagine that each individual produces a large number of gametes, that the males of the next generation are obtained by the random union of $M$ male gametes with $M$ female gametes, and likewise for the $F$ females of the next generation.

Conditional on the gene frequencies in generation $t$, $X_{t+1}^*$ is the sum of two independent binomial variates, both with index $M$ and with probabilities $P_t^*$ and $P_t^{**}$; likewise $X_t^{**}$ is the sum of two independent binomial variates with index $F$ and with probabilities $P_t^*$ and $P_t^{**}$. Thus we may write

$$P_{t+1}^* = \tfrac{1}{2}(P_t^* + P_t^{**}) + \mathrm{d}P^*$$
$$P_{t+1}^{**} = \tfrac{1}{2}(P_t^* + P_t^{**}) + \mathrm{d}P^{**}$$

(12.18)

where $\mathrm{d}P^*$ and $\mathrm{d}P^{**}$ are independent random variables with zero means and with variances

$$\mathrm{Var}(\mathrm{d}P^*) = [P_t^*(1 - P_t^*) + P_t^{**}(1 - P_t^{**})]/4M$$
$$\mathrm{Var}(\mathrm{d}P^{**}) = [P_t^*(1 - P_t^*) + P_t^{**}(1 - P_t^{**})]/4F.$$

(12.19)

To simplify the analysis of this system we consider the variables

$$P_t = \tfrac{1}{2}(P_t^* + P_t^{**})$$
$$Y_t = \tfrac{1}{2}(P_t^* - P_t^{**}).$$

(12.20)

Conditional on $P_t$ and $Y_t$ we find that

$$P_{t+1} = P_t + \mathrm{d}P$$
$$Y_{t+1} = \mathrm{d}Y$$

(12.21)

where $\mathrm{d}P$ and $\mathrm{d}Y$ are uncorrelated random variables with zero mean and with the same variance

$$\mathrm{Var}(\mathrm{d}P) = \mathrm{Var}(\mathrm{d}Y) = [P_t(1 - P_t) - Y_t^2]\left[\frac{1}{8M} + \frac{1}{8F}\right].$$

(12.22)

We observe that

$$E(Y_t^2 | P_{t-1}, Y_{t-1}) = [P_{t-1}(1 - P_{t-1}) - Y_{t-1}^2]\left[\frac{1}{8M} + \frac{1}{8F}\right]$$

(12.23)

$$\leqslant P_{t-1}(1 - P_{t-1})\left[\frac{1}{8M} + \frac{1}{8F}\right].$$

Thus if $M$ and $F$ are reasonably large, $Y_t^2$ will be small compared with $P_{t-1}(1 - P_{t-1})$; furthermore the latter quantity will be approximately the same as $P_t(1 - P_t)$ since $P_t$ changes rather slowly when the population size is large. Hence

$$\mathrm{Var}(\mathrm{d}P) = \mathrm{Var}(\mathrm{d}Y) \simeq P_t(1 - P_t)\left[\frac{1}{8M} + \frac{1}{8F}\right].$$

(12.24)

We now define the effective population size, $N_e$, as

$$N_e = \left[ \frac{1}{4M} + \frac{1}{4F} \right]^{-1} \tag{12.25}$$

Thus

$$\mathrm{Var}(dP) \simeq P_t(1 - P_t)/2N_e \tag{12.26}$$

From eqns (11.21) and (11.26) we see that $P_t$ as defined in eqn (11.20) behaves in the same way as the gene frequency without separate sexes with size $N = N_e$. If the numbers of males and females are approximately the same, then $N_e \simeq (M + F)$, as one would expect. If, however, there are many more females than males, then $N_e \simeq 4M$; this situation arises in polygynous species (for example in elephant seals each breeding bull has a large harem of females) and in some farm animals (for example in a dairy herd only a very few selected bulls are used for breeding).

## The effect of genetic drift on quantitative characters

Random fluctuations due to finite population size can affect the mean and variance of a quantitative character in two ways, through random drift of gene frequencies and through random departures from Hardy–Weinberg and linkage equilibrium. Changes in gene frequencies are permanent, whereas departures from equilibrium are only temporary and short-lived. We shall first consider the effect of gene frequency drift; departures from equilibrium will be considered separately in the next section. We assume a random mating population of size $N$ without separate sexes. We suppose that the existence of separate sexes and other possible complications can be adequately taken into account by an appropriate definition of effective population size.

We first consider a single locus with two alleles, $B_1$ and $B_2$, without dominance. Suppose that $B_1$ contributes $a$ to the character value and $B_2$ nothing, so that the contributions of the genotypes $B_1B_1$, $B_1B_2$, and $B_2B_2$ are $2a$, $a$, and $0$ respectively. If the frequency of $B_1$ in generation $t$ is $P_t$, then the contribution to the mean from this locus is

$$M_t = 2aP_t. \tag{12.27}$$

If the population were in perfect Hardy–Weinberg and linkage equilibrium, the contribution to the variance from this locus would be

$$V_t = 2a^2 P_t(1 - P_t). \tag{12.28}$$

We therefore *define* this quantity as the contribution to the 'true' genetic variance at this locus.

Consider the effect of genetic drift on the mean value, $M_t$. From eqn (12.5) we find that

$$E(M_t) = M_0$$

$$\mathrm{Var}(M_t) = 2V_0 \left[ 1 - \left( 1 - \frac{1}{2N} \right)^t \right]. \tag{12.29}$$

When $t \ll 2N$,

$$\mathrm{Var}(M_t) \simeq V_0 t/N. \tag{12.30}$$

Thus the mean value of the character remains, on average, unchanged but may vary between lines (that is to say, from one realization of the process to the next).

The above results remain valid for multiple alleles, $B_1$, $B_2$, ..., $B_k$ with additive effects $a_1, a_2, \ldots, a_k$. We define

$$\begin{aligned} M_t &= 2\sum a_j P_j(t) \\ V_t &= 2\sum a_j^2 P_j(t) - 2(\sum a_j P_j(t))^2. \end{aligned} \tag{12.31}$$

It follows from eqns (12.5) and (12.15) that eqn (12.29) holds. The results can also be generalized to multiple loci. Define $M_i(t)$ and $V_i(t)$ as the contributions to the mean and the 'true' variance from the $i$th locus, and define

$$\begin{aligned} M_t &= \sum M_i(t) \\ V_t &= \sum V_i(t) \end{aligned} \tag{12.32}$$

It is clear that eqn (12.29) still holds if we assume that the gene frequencies at different loci are drifting independently of each other; this assumption will be justified in the next section.

It can be concluded that, with additive gene action, the variance of the 'true' mean due to genetic drift is $\mathrm{Var}(M_t) \simeq V_A t/N$, where $V_A$ is the additive genetic variance in the base population. This is the variability which would be observed between different replicate lines, each of effective size $N$. It has been assumed that there is no selection, but Hill (1977) has suggested that the same result holds approximately under selection if $N$ is interpreted as the effective number of breeding individuals. This is important in the interpretation of artificial selection experiments which are usually done on rather small numbers of animals. There may well be appreciable differences between replicate experiments due to genetic drift, and erroneous conclusions may be drawn if this source of variability is not taken into account. Falconer (1977) discusses the results of selection experiments on mice, and shows that apparently significant differences between lines selected for high and low body weight may be largely due to drift.

We turn now to the effect of drift on the variance, $V_t$. It is easily verified from the results of the previous section that

$$E(V_t) = \left(1 - \frac{1}{2N}\right)^t V_0.$$  (12.33)

This is true for an arbitrary number of alleles or loci. Thus the 'true' variance declines on average at a rate of $1/2N$ per generation. It is not possible to make any useful general statement about $\mathrm{Var}(V_t)$ except that it is inversely proportional to the number of loci. (From dimensional considerations $\mathrm{Var}(V_i(t))$ is of order $V_i^2(0)$. As we increase the number of loci keeping the total variance constant $V_i(0)$ must be of order $1/n$, so that $\mathrm{Var}(\sum V_i(t)) = \sum \mathrm{Var}(V_i(t))$ is of order $1/n$.) If we assume a single locus with two alleles so that $V_t$ is given by eqn (12.28), then

$$\mathrm{Var}(V_t) = a^4 \, \mathrm{Var}(H_t),$$  (12.34)

where $\mathrm{Var}(H_t)$ is given by eqn (12.42). For $n$ loci with two alleles with equal effects and with the same initial gene frequencies we find that

$$\mathrm{Var}(V_t) = \frac{V_0^2}{n}\left\{\frac{\left(1 - \dfrac{1}{2N}\right)^t}{5pq} + \frac{(5pq - 1)}{5pq}\left(1 - \frac{3}{N}\right)^t - \left(1 - \frac{1}{2N}\right)^{2t}\right\}.$$  (12.35)

The standard error of the change in the variance may be appreciable compared with the predicted mean change.

We shall now investigate the effect of dominance. Consider a locus with two alleles, and suppose that the contributions of the genotypes $B_1 B_1$, $B_1 B_2$, and $B_2 B_2$ are $a$, $d$, and $-a$ respectively, If the frequency of $B_1$ in generation $t$ is $P_t$, the contribution of this locus to the 'true' mean (on the assumption of Hardy–Weinberg and linkage equilibrium) is

$$M_t = -(1 - 2P_t)a + 2P_t(1 - P_t)d.$$  (12.36)

From the results of the previous section we find that

$$E(M_t) = M_0 - \left[1 - \left(1 - \frac{1}{2N}\right)^t\right]2pq\,d$$

$$\mathrm{Var}(M_t) = 4a^2\,pq\left[1 - \left(1 - \frac{1}{2N}\right)^t\right]$$

$$- 4adpq(p - q)\left\{\left(1 - \frac{1}{2N}\right)^t - \left(1 - \frac{3}{2N}\right)^t\right\}$$

$$+ d^2\,\mathrm{Var}(H_t).$$  (12.37)

The result for multiple loci can be obtained by summing over loci. Thus there may be a change in the predicted mean value due to drift in the presence of dominance. This may be regarded as a manifestation of inbreeding depression due to the inbreeding effect of finite population size. However, the magnitude of the Expected change may be small compared with its standard error unless the number of loci is large.

To simplify the analysis in considering the change in the variance we shall suppose that $B_1$ is completely dominant to $B_2$, so that the effects of the genotypes $B_1B_1$, $B_1B_2$, and $B_2B_2$ are $a$, $a$, and $-a$ respectively. The contribution of this locus to the 'true' variance is

$$V_t = 4P_t(1 - P_t)^2 (2 - P_t)a^2. \tag{12.38}$$

This contribution may be partitioned into additive and dominance components:

$$V_A(t) = 8P_t(1 - P_t)^3 a^2$$
$$V_D(t) = 4P_t^2(1 - P_t)^2 a^2. \tag{12.39}$$

From the results of the previous section we find that

$$E(V_t) = \left\{ \tfrac{4}{3}pq \left(1 - \frac{1}{2N}\right)^t + pq(q - p)\left(1 - \frac{3}{2N}\right)^t \right.$$
$$\left. - pq(pq - \tfrac{1}{3})\left(1 - \frac{3}{N}\right)^t \right\}4a^2$$

$$E(V_A(t)) = \left\{ \tfrac{2}{3}pq\left(1 - \frac{1}{2N}\right)^t + pq(q - p)\left(1 - \frac{3}{2N}\right)^t \right.$$
$$\left. - 2pq(pq - \tfrac{1}{3})\left(1 - \frac{3}{N}\right)^t \right\}4a^2 \tag{12.40}$$

$$E(V_D(t)) = \left\{ \tfrac{1}{3}pq\left(1 - \frac{1}{2N}\right)^t + pq(pq - \tfrac{1}{3})\left(1 - \frac{3}{N}\right)^t \right\}4a^2.$$

These quantities eventually decline to zero, but they may show an initial increase. The total contribution shows an initial increase, that is to say $E(V_1) > V_0$, when $q < 0.41$; the additive component shows an initial increase when $q < \tfrac{1}{2}$, and the dominance component when $pq < \tfrac{1}{6}$.

### *Random departures from Hardy–Weinberg and linkage equilibrium

The second way in which stochastic fluctuations due to finite population size manifest themselves is in departures from Hardy–Weinberg and linkage equilibrium. Consider a character determined by $n$ loci without dominance or epistasis in a population of size $N$. The genotypic value of the $j$th individual in generation $t$ can be represented as

$$G_j(t) = \sum_i (X_{ij}(t) + X'_{ij}(t)) \tag{12.41}$$

where $X_{ij}$ and $X'_{ij}$ are the contributions at the $i$th locus from the paternal and maternal gametes. Define

$$X_{i.} = \sum_j (X_{ij} + X'_{ij})/2N$$

$$d_{ij} = X_{ij} - X_{i.}, \quad d'_{ij} = X'_{ij} - X_{i.}. \tag{12.42}$$

The actual genotypic mean and variance in the population are

$$M = \text{Genotypic mean} = 2\sum_i X_{i.}$$

$$V = \text{Genotypic variance} = \sum_j [\sum_i (d_{ij} + d'_{ij})]^2/N. \tag{12.43}$$

Consider first the genotypic mean. $X_{i.}$ is a linear function of the gene frequencies at the $i$th locus; it is therefore affected by genetic drift in the gene frequencies as discussed in the previous section but is unaffected by departures from Hardy–Weinberg equilibrium. Furthermore, the covariance of $X_{i.}$ and $X_{k.}$, for $i \neq k$, must be zero from symmetry considerations since positive and negative effects of disequilibrium are equally likely (i.e. any pair of loci must *on average* be in equilibrium). Hence neither $E(M)$ nor $\text{Var}(M)$ is affected by departures from Hardy–Weinberg or linkage equilibrium.

By contrast the genotypic variance is directly affected by disequilibrium. It is useful to decompose the variance into components representing the effects of gene frequencies and of departures from Hardy–Weinberg and linkage equilibrium. To do this we define

$$C_{ik} = \sum_j (d_{ij} d_{kj} + d'_{ij} d'_{kj})/N$$

$$C^*_{ik} = \sum_j (d_{ij} d'_{kj} + d'_{ij} d_{kj})/N. \tag{12.44}$$

It is clear that

$$V = \sum_{i,k} (C_{ik} + C^*_{ik}). \tag{12.45}$$

$C_{ii}$ represents the 'true' genetic variance contributed by the $i$th locus, and $\sum C_{ii}$ the total 'true' genetic variance, as discussed in the previous section. $C^*_{ii}$ represents the effect of departure from Hardy–Weinberg equilibrium at the $i$th locus, $C_{ij}$ ($i \neq k$) represents the effect of linkage disequilibrium between loci $i$

and $k$, and $C_{ik}^{*}$ represents the effect of departure from Hardy–Weinberg equilibrium between loci $i$ and $k$. (Note that the $C_{ik}$s have the character of inter-class covariances based on $2N$ observations, while the $C_{ik}^{*}$s have the character of intra-class covariances based on $N$ pairs of observations.)

We shall now extend the 'random union of gametes' model to this situation. Given the population as observed in the $t$th generation, we suppose that it forms an effectively infinite pool of gametes determining a vector-valued random variable $\mathbf{X} = (X_1, X_2, \ldots, X_n)$, where $X_i$ is the contribution which a gamete would make to the next generation at the $i$th locus. The moments of this gametic distribution, conditional on the parental population, are

$$E(X_i) = X_i(t)$$

$$\text{Var}(X_i) = \tfrac{1}{2}C_{ii}(t) \tag{12.46}$$

$$\text{Cov}(X_i, X_k) = \tfrac{1}{2}[(1 - r_{ik})C_{ik}(t) + r_{ik}\,C_{ik}^{*}(t)],$$

where $r_{ik}$ is the recombination fraction between loci $i$ and $k$. (The factor $\tfrac{1}{2}$ allows for the reduction from the diploid to the haploid state; the covariance term also allows for recombination.) The next generation is then formed by selecting $2N$ gametes at random and then pairing them at random. Thus conditional on the $X_{ij}(t)$ and $X_{ij}'(t)$, $X_{ij}(t + 1)$ has Expected value and variance given by eqn (12.46), and $X_{ij}(t + 1)$ and $X_{kj}(t + 1)$ have the covariance given by (12.46); on the other hand, $X_{ij}(t + 1)$ and $X_{kj}'(t + 1)$ are uncorrelated for all $i$ and $k$, as are $X_{ij}(t + 1)$ and $X_{kl}(t + 1)$ for $j \neq l$.

By writing out the definitions of $C_{ik}(t + 1)$ and $C_{ik}^{*}(t + 1)$ and taking Expected values conditional on the previous generation, we find that

$$E(C_{ij}(t + 1)|t) = \left(1 - \frac{1}{2N}\right)C_{ii}(t)$$

$$E(C_{ii}^{*}(t + 1)|t) = -C_{ii}(t)/2N$$

$$E(C_{ik}(t + 1)|t) = \left(1 - \frac{1}{2N}\right)[(1 - r_{ik})C_{ik}(t) \tag{12.47}$$
$$+ r_{ik}\,C_{ik}^{*}(t)]$$

$$E(C_{ik}^{*}(t + 1)|t) = -C_{ik}(t)/2N.$$

(To evaluate the Expected values, observe that

$$d_{ij}(t + 1) = X_{ij}(t + 1) - X_{i.}(t + 1) = (X_{ij}(t + 1) - X_{i.}(t))$$
$$- (X_{i.}(t + 1) - X_{i.}(t))$$

and hence find $E(d_{ij}(t + 1)d_{kj}(t + 1))$ and so on.)

The variances and covariances of the $C_{ik}$s and $C_{ik}^{*}$s can be found in a similar way. In this case it is only necessary to keep terms of order $1/N$; we can thus to sufficient accuracy write $d_{ij}(t + 1) = X_{ij}(t + 1) - X_{i.}(t)$. We shall also assume that $C_{ik}$ and $C_{ik}^*$ are negligible in comparison with $C_{ii}$ and $C_{kk}$ when $i \neq k$; this assumption will be justified by the results in eqn (12.55) provided that $r_{ik}N$ is large. With these assumptions the terms $d_{ij}(t + 1)$ conditional on generation $t$ can all be regarded as independent random variables with zero mean and variance $\frac{1}{2}C_{ii}(t)$. The following variances, correct to order $1/N$, can now be computed in a straightforward way:

$$
\begin{aligned}
\mathrm{Var}(C_{ii}(t + 1)|t) &= [4m_{4i}(t) - C_{ii}^2(t)]/2N \\
\mathrm{Var}(C_{ik}(t + 1)|t) &= C_{ii}(t)C_{kk}(t)/2N, \quad i \neq k \\
\mathrm{Var}(C_{ii}^*(t + 1)|t) &= C_{ii}^2(t)/N \\
\mathrm{Var}(C_{ik}^*(t + 1)|t) &= C_{ii}(t)C_{kk}(t)/2N, \quad i \neq k.
\end{aligned}
\tag{12.48}
$$

$m_{4i}(t)$ is the fourth moment of the gametic contribution at the $i$th locus. The corresponding covariances are all zero to the same order of approximation.

We first consider the behaviour of the 'true' variance $C_{ii}$. Taking Expected values in the first equation in (12.47) conditional only on $C_{ii}(0)$ we find that

$$
E(C_{ii}(t)) = \left(1 - \frac{1}{2N}\right)^t C_{ii}(0).
\tag{12.49}
$$

This agrees with eqn (11.33). To obtain information about the variance of $C_{ii}$ we must make some assumption about $m_{4i}$. With two alleles this problem has been solved in eqn (12.35) (with $n = 1$). With a large number of alleles, it seems likely that the distribution of effects is normal so that $4m_{4i}(t) = 3C_{ii}^2(t)$. In this case

$$
E(C_{ii}^2(t + 1)|C_{ii}(t)) = C_{ii}^2(t) + o(1/N)
$$

so that

$$
E(C_{ii}^2(t)|C_{ii}(0)) = C_{ii}^2(0)
$$

$$
\mathrm{Var}(C_{ii}(t)) = \left[1 - \left(1 - \frac{1}{2N}\right)^{2t}\right]C_{ii}^2(0).
\tag{12.50}
$$

However, the expression for the variance does not tend to zero, as it ought, when $t$ becomes large; it is likely that the assumption of normality breaks down as the variance declines.

We now regard $C_{ii}(t)$ as constant over the short term, since from the first line in eqn (12.48) its coefficient of variation is of order $N^{-1/2}$, and we consider the behaviour of the other terms. We neglect terms of order $1/N$ in eqn (12.47), which are negligible since the variance in eqn (12.48) is of order $1/N$. Hence the Hardy–Weinberg disequilibrium terms, $C_{ik}^*$, behave like a completely random process, uncorrelated in time, with zero mean and with variances given by eqn

(12.48). Denote the sum of these terms by $C_{HW}$, the contribution of departures from Hardy–Weinberg equilibrium to the observed variance:

$$C_{HW} = \sum_{i,k} C_{ik}^* = 2 \sum_{i<k} C_{ik}^* + \sum_i C_{ii}^*. \tag{12.51}$$

Thus $C_{HW}$ is a completely random process with zero Expectation and with variance

$$\mathrm{Var}(C_{HW}) = 2 \sum_{i<k} C_{ii} C_{kk}/N + \sum_i C_{ii}^2/N = (\sum_i C_{ii})^2/N = V_T^2/N. \tag{12.52}$$

Thus departures from Hardy–Weinberg equilibrium cause random fluctuations of the observed genetic variance about the 'true' genetic variance, $V_T$; these fluctuations have Expected root mean square $V_T/N^{1/2}$, and they are uncorrelated from one generation to the next.

We turn now to the linkage disequilibrium term, $C_{ik}$, which behaves effectively, for $i \neq k$, like a first order autoregressive process

$$C_{ik}(t+1) = (1 - r_{ik})C_{ik}(t) + e_t, \tag{12.53}$$

where the $e_t$s are uncorrelated 'errors' with zero mean and with variance

$$\mathrm{Var}(e_t) = (1 + r_{ik}^2)C_{ii} C_{kk}/2N. \tag{12.54}$$

The steady-state variance of $C_{ik}$ is

$$\mathrm{Var}(C_{ik}) = \frac{(1 + r_{ik}^2)}{r_{ik}(2 - r_{ik})} \frac{C_{ii} C_{kk}}{2N}. \tag{12.55}$$

There is a correlation $(1 - r_{ik})$ between successive values of $C_{ik}$. For unlinked loci

$$\mathrm{Var}(C_{ik}) = 5C_{ii} C_{kk}/6N. \tag{12.56}$$

Denote the sum of these terms by $C_L$, the contribution to the observed variance due to linkage disequilibrium:

$$C_L = 2 \sum_{i<k} C_{ik}. \tag{12.57}$$

In the absence of linkage

$$\mathrm{var}(C_L) = 10 \sum_{i<k} C_{ii} C_{kk}/3N = \frac{5}{3N}\left[(\sum C_{ii})^2 - \sum C_{ii}^2\right] \simeq 5V_T^2/3N. \tag{12.58}$$

Thus departures from linkage equilibrium cause random fluctuations of the observed variance about the 'true' variance; in the absence of linkage these fluctuations have root mean square $\sqrt{(5/3N)}V_T$ and there is a correlation of $\frac{1}{2}$ between successive values, but both the root mean square and the correlation increase under linkage.

To summarize our conclusions, we may represent the observed variance as the sum of three components:

$$V = V_T + C_{HW} + C_L. \tag{12.59}$$

In the decomposition, $V_T$ is the 'true' variance, which would be observed given the observed gene frequencies if there were perfect Hardy–Weinberg and linkage equilibrium, and $C_{HW}$ and $C_L$ are fluctuations of the observed about the true variance due to departures from Hardy–Weinberg and linkage equilibrium. $V_T$ declines at an average rate of $1/2N$ per generation, but over a time span which is short compared to the population size it can be regarded as nearly constant. $C_{HW}$ and $C_L$ cause rapid fluctuations about $V_T$ which are on average zero but which have mean square deviations of $V_T^2/N$ and $5V_T^2/3N$ respectively; the latter value is increased by linkage.

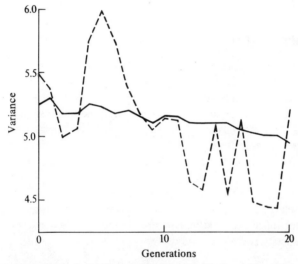

Fig. 12.1. 'True' variance (solid line) and observed genetic variance (dotted line) in a computer simulation (Bulmer 1976b).

Fig. 12.1 shows a computer simulation which illustrates these points (Bulmer 1976b). In this simulation the character value was determined by 12 unlinked loci with additive gene action, and there were 500 individuals of each sex in each generation, of whom 100 of each sex were chosen at random as the parents of the next generation. As predicted there is a gradual decline in the 'true' variance (the solid line) at a rate of about $1/2N$ per generation, where $N = 200$, the size of the breeding population. It can also be seen that the observed genetic variance (the dotted line) fluctuates quite noticeably above and below the 'true' variance, due to departures from Hardy–Weinberg and linkage equilibrium. The magnitude of these fluctuations is rather smaller than would be predicted from the foregoing theory with $N = 200$, and is consistent with an effective population size for this purpose of about 350; these fluctuations are presumably affected not only by the size of the breeding population (200) but also by the size of the total population from which the variance was calculated (1000). The theory developed here does not take this distinction into account; it seems likely that the size of the fluctuations is quite sensitive to the breeding structure of the population. It should also be emphasized that the present theory does not take dominance into account. Similar (though not identical) theoretical results have been obtained by Bulmer (1976b) and Avery and Hill (1977). Insofar as previous results differ from the present results, I believe the former to be incorrect, though the problem is clearly a difficult one. Avery and Hill (1977) ignore departures from Hardy–Weinberg equilibrium. They thus miss the important term $\text{Var}(C_{\text{HW}})$, and they omit the term $r_{ik}^2$ in eqns (12.54) and (12.55) which comes from the term $R_{ik}C_{ik}^*(t)$ in the third line of eqn (12.47); hence they find that, in the absence of linkage, $\text{Var}(C_{\text{L}}) = 4V_T^2/3N$ rather than $5V_T^2/3N$.

## Limits to artificial selection

In a large population, selection may be expected to increase the frequency of favourable alleles, which increase the value of the character being selected, until they eventually reach fixation. In a small population, however, some of these favourable alleles may be lost by chance, so that the total response to selection may be less than the maximum response possible in a large population. This problem has been considered by Robertson (1960, 1970). We shall here discuss the theory developed in the earlier paper, which ignores complications due to linkage.

We shall first derive the chance of fixation of a specified allele under the joint effects of selection and genetic drift by using the diffusion approximation approach; for fuller details see Crow and Kimura (1970, Chapter 8). Consider a single locus with two alleles and write

$$P_{t+1} = P_t + \mathrm{d}P \qquad (12.1\ bis)$$

where $\mathrm{d}P$ is the change in the frequency of one of the alleles in one generation as a result of selection and of sampling fluctuations. Write

$$E(\mathrm{d}P) = M(P_t)$$
$$\text{Var}(\mathrm{d}P) = V(P_t) \qquad (12.60)$$

$M$ is the mean change in gene frequency due to selection; $V$ is the variance in the change due to random sampling fluctuations. It will be assumed that changes in different generations are independent of each other, apart from their dependence on the current gene frequency.

To a good approximation we may regard both the gene frequency $P_t$ and time as continuous variables. Let

$$\mathrm{d}P = P_{t+\mathrm{d}t} - P_t \qquad (12.61)$$

be the change in gene frequency in a short time interval $dt$. We suppose that

$$E(dP) = M(P_t)\, dt$$
$$\text{Var}(dP) = V(P_t)\, dt \tag{12.62}$$

and that changes in non-overlapping time intervals are independent. We denote by $f(p, x; t)$ the probability density that $P_t$ lies between $x$ and $x + dx$ at time $t$, given that $P_0 = p$. We shall now show that $f$ satisfies the Kolmogorov backward equation:

$$\frac{\partial f(p, x; t)}{\partial t} = M(p)\frac{\partial}{\partial p}f(p, x; t) + \tfrac{1}{2}V(p)\frac{\partial^2}{\partial p^2}f(p, x; t). \tag{12.63}$$

To derive this result we observe that

$$f(p, x; t + dt) = \int f(p, y; dt)f(p + y, x; t)\, dy. \tag{12.64}$$

This equation states that to go from $p$ to $x$ in time $t + dt$ we must go from $p$ to $p + y$ in time $dt$ and then from $p + y$ to $x$ in time $t$, for some $y$; the two events are independent so that their probabilities are to be multiplied together and then summed over all possible values of $y$. Expanding both sides in Taylor series and ignoring powers of $dt$ and powers of $y$ higher than $y^2$, we find that

$$f + \frac{\partial f}{\partial t}dt = \int f(p, y; dt)\left\{f + y\frac{\partial f}{\partial p} + \tfrac{1}{2}y^2\frac{\partial^2 f}{\partial p^2}\right\}dy \tag{12.65}$$

where $f$ means $f(p, x; t)$ unless otherwise stated. Eqn (12.63) follows immediately by observing that $y$ is the change in the gene frequency in the initial time interval $dt$, whose mean and variance are given by eqn (12.62) with $P_t = p$.

We now denote by $u(p)$ the probability that an allele will ultimately be fixed rather than lost, starting from the initial gene frequency $p$. This probability can be obtained from $f(p, x; t)$ by putting $x = 1$ and $t = \infty$. Since $\partial f/\partial t = 0$ at $t = \infty$, we find from eqn (12.63) that $u(p)$ satisfies the equation

$$M(p)\frac{du(p)}{dp} + \tfrac{1}{2}V(p)\frac{d^2 u(p)}{dp^2} = 0 \tag{12.66}$$

with boundary conditions

$$u(0) = 0$$
$$u(1) = 1. \tag{12.67}$$

Eqn (12.66) may be written in the form

$$\frac{d}{dp}\left(\ln\frac{du(p)}{dp}\right) = -\frac{2M(p)}{V(p)}. \tag{12.68}$$

The solution which satisfies the boundary conditions is

$$u(p) = \int_0^p G(x)\,dx \Big/ \int_0^1 G(x)\,dx \tag{12.69}$$

where

$$G(x) = \exp - \int \frac{2M(x)}{V(x)}\,dx. \tag{12.70}$$

(The integral in eqn (12.70) is an indefinite integral; the constant of integration can be chosen arbitrarily without affecting the final result.)

We now consider the simplest case in which the fitnesses of the three genotypes $B_1B_1$, $B_1B_2$, and $B_2B_2$ are $1 + s$, $1$, and $1 - s$ respectively; we are thus assuming that $B_1$ has a constant advantage over $B_2$ without any dominance in fitness. The mean change in gene frequency in one generation of selection, if the frequency of $B_1$ is $P_t = x$, is

$$M(x) = sx(1 - x). \tag{12.71}$$

The variance of the change due to sampling errors is assumed to be the same as in the absence of selection:

$$V(x) = x(1 - x)/2N, \tag{12.72}$$

so that $2M(x)/V(x) = 4Ns$. Hence

$$u(p) = \frac{1 - e^{-4Nsp}}{1 - e^{-4Ns}}. \tag{12.73}$$

We now turn to the effect of directional selection on a quantitative character. Suppose that there are $n$ loci each with two alleles, say $B_{1i}$ and $B_{2i}$ at the $i$th locus, and that $B_{1i}$ contributes $a_i$ to the character value and $B_{2i}$ nothing, without dominance or epistasis. By comparing eqns (12.10–19) with eqn (12.71) we find that the selection coefficient acting at the $i$th locus is

$$s_i = DMa_i/V, \tag{12.74}$$

where $DM$ is the change in the mean as a direct result of selection and $V$ is the phenotypic variance. Under truncation selection with proportion selected $\pi$, $DM/V = IV^{1/2}$, where $I = \phi(z)/\pi$, so that

$$s_i = IV^{-1/2}\,a_i. \tag{12.75}$$

Strictly speaking eqn (12.75) is only valid in the first generation of selection; the selection coefficient will change subsequently, in the short term due to the build-up of linkage disequilibrium and in the long term due to changes in gene frequencies. However, we shall ignore these changes and suppose that selection acts over many generations with constant selection coefficients at individual loci given by eqn (12.75). Eventually all loci will become fixed at either $B_{1i}$ or

$B_{2i}$, with probabilities given by eqn (12.73), and the population will cease to respond to selection. When this selection limit is reached the average change in the mean value of the character is given by

$$\Delta M = \sum_i 2a_i [u(p_i) - p_i] = \sum_i 2a_i \frac{[1 - e^{-4NS_ip_i} - p_i + p_i e^{-4NS_i}]}{1 - e^{-4NS_i}}. \quad (12.76)$$

When $Ns_i$ is large, $u(p_i) \simeq 1$; the total response to selection in a very large population is

$$\Delta M = \sum_i 2a_i(1 - p_i). \quad (12.77)$$

When $Ns_i$ is small, we may expand the exponentials in power series, and obtain the approximate result

$$\Delta M = \sum_i 4Na_i s_i p_i(1 - p_i) = 2NIV^{-1/2} \sum_i 2a_i^2 p_i(1 - p_i)$$

$$= 2NIV^{-1/2} V_G, \quad (12.78)$$

where $V_G$ is the initial genetic variance and $V$ is the initial phenotypic variance. Thus the total predicted response is $2N$ times the predicted response in the first generation.

This result can be used to find the optimum intensity of selection. Consider a selection programme in which $T$ animals are measured in each generation and a proportion $\pi$ is selected with the highest values for some metric character. How should we choose $\pi$ to maximize the total predicted response (12.78)? Putting $N = T\pi$, $I = \phi(z)/\pi$, we find that $NI = T\phi(z)$. But $\phi(z)$, which is the standard normal density function, has a maximum at $z = 0$, $\pi = \frac{1}{2}$. Thus to obtain the largest total response after many generations we should select 50 per cent of the population in each generation.

So far we have discussed the ultimate selection limit, but we have not considered the time taken to reach this limit. We shall now find the 'half-life' of the selection process, defined as the time taken for the average value of the mean to reach halfway to the eventual selection limit. It is possible to obtain a simple approximate solution to this difficult problem under weak selection. Consider a single locus with selection coefficient $s$ acting in favour of the $B_1$ allele, and suppose that $Ns$ is small. From eqn (12.71) the mean change in the gene frequency due to selection in generation $t$ is $sP_t(1 - P_t)$. Since selection is weak we may suppose that the mean heterozygosity declines at a rate of $1/2N$ per generation, as it does in the absence of selection. Hence

$$E[P_t(1 - P_t)|P_0 = p] \simeq \left(1 - \frac{1}{2N}\right)^t p(1 - p) \quad (12.8\ bis)$$

so that

$$E(P_t|P_0=p) = p + \sum_{\tau=0}^{t-1} sE[P_\tau(1-P_\tau)|P_0=p]$$

$$\simeq p + sp(1-p) \sum_{\tau=0}^{t-1} \left(1-\frac{1}{2N}\right)^\tau$$

$$\simeq p + 2Nsp(1-p)(1-e^{-t/2N}).\qquad(12.79)$$

The average gene frequency increases from $p$ at $t = 0$ to $p + 2Nsp(1 - p)$ at $t = \infty$. The half-life is the time at which the average gene frequency has travelled half this way, which is the value of $t$ satisfying $e^{-t/2N} = \frac{1}{2}$; thus

$$\text{half-life} = 1.4N \text{ generations.}\qquad(12.80)$$

This is also the half-life for the mean value of the character, which depends linearly on the gene frequencies.

Thus when $Ns$ is small for all loci, the selection limit can be predicted from eqn (12.78), which can be calculated from quantities measurable in the base population, and the half-life of the process is $1.4N$ generations. This theory is applicable to characters determined by many loci with individually small effects, but little can be said in general terms about characters determined by a few loci with large effects for which $Ns$ cannot be assumed small at individual loci. For example, eqn (12.77) gives the total response to selection in a very large population, but it involves the individual gene effects, $a_i$, and gene frequencies, $p_i$, which cannot be measured in the base population. It seems likely, however, that both the total response to selection and the half-life of the process will be less for a character determined by a small number of loci than the limiting results for weak selection in eqns (12.78) and (12.80).

It must also be remembered that the above theory rests on a number of simplifying assumptions and can thus only be regarded as approximate. The main assumptions are as follows: (i) additive gene action; (ii) no linkage; (iii) that the selection coefficient $s$ remains constant throughout the selection process; (iv) two alleles per locus; (v) that the effective population size in artificial selection is equal to the number of parents selected. In connection with the last assumption, it seems likely that the effective population size is less than $N$, and that the discrepancy increases with the selection intensity, since the parents selected will not contribute equally to future generations; the parents with a higher breeding value will have a higher chance of contributing than those with a lower breeding value. For further discussion of these and other points see Robertson (1960, 1970). The theory has been evaluated in the light of experimental evidence by Frankham (1977b).

## Problems

1. A strain of mice is maintained in the laboratory by choosing five males and ten females in each generation at random to be the parents of the next generation. What is the effective population size? What would you expect the mean litter size to be, on average, after ten generations if the base population is the random-bred population of Problem 1, Chapter 7?

2. Suppose that two replicate lines of mice are maintained by the above procedure. Consider a character with additive gene action and with $m = V_Y = 100$, $h^2 = \frac{1}{2}$ in the base population. Suppose that the phenotypic values of the 15 breeding individuals in each line are measured after ten generations. Find the predicted variance of the difference between the mean values of the two lines, and hence find the probability that they will differ by ten units or more.

3. Consider a locus with two alleles, $B_1$ and $B_2$, with a selection coefficient $s = 0.01$ acting in favour of $B_1$. Find the probability that this allele will eventually be fixed for the following values of $N$ and $p$: $N = 20, 100, 500; p = 0.1, 0.5, 0.9$.

4. Consider a character determined by a large number of loci with additive gene action and with $m = V_Y = 100$, $h^2 = \frac{1}{2}$ in the base population. If the top ten out of 50 measured individuals are selected in each generation, what is the predicted mean value when response to selection ceases? After how many generations would you expect 75 per cent of this response to be achieved?

# Answers to problems

## Chapter 1

1. (a) $\frac{1}{2}$ red, $\frac{1}{2}$ pink; (b) $\frac{1}{2}$ white, $\frac{1}{2}$ pink.

2. (a) Heterozygous; (b) either heterozygous or homozygous.

3. Chi square values: 0.59 with 2 d.f. for independence, 0.56 with 2 d.f. for colour segregation, 0.14 with 1 d.f. for shape segregation.

4. (second part) $\frac{1}{2}$ purple, $\frac{1}{2}$ white or $\frac{1}{4}$ purple, $\frac{1}{4}$ red, $\frac{1}{2}$ white depending on which white variety is used.

5. Chi square values, each with 1 d.f.: 3393 for independence, 0.04 for colour, segregation, 0.08 for pollen shape segregation. Expected proportions are $\frac{1}{4}(2 + t)$, $\frac{1}{4}(1 - t)$, $\frac{1}{4}(1 - t)$, and $\frac{1}{4}t$, where $t = (1 - r)^2$; predicted numbers with $r = 0.12$ are 4822, 392, 392, and 1346.

6. (first part) all red; (second part) (a) 1, (b) $\frac{1}{2}$.

7. Females: $\frac{1}{2}$ black, $\frac{1}{2}$ tortoiseshell; males: $\frac{1}{2}$ black, $\frac{1}{2}$ ginger.

8. Binomial with probability $\frac{1}{2}$ and index $n$. The mean will be 53.6 and the variance 23.0 (estimating the environmental variance as $\frac{1}{2}(2.3 + 7.9) = 5.1$).

9. Slope = 0.78. Because genotypic values of $F_2$ parents will on average be less extreme than their phenotypic values.

## Chapter 2

1. $V_Y = 9681$, $V_G = 3883$, $V_E = 5798$.

2. The standard deviation is approximately proportional to the mean; a logarithmic transformation will stabilize the variance.

3.

| Source of variation | Sum of squares | Degrees of freedom | Mean square |
|---|---|---|---|
| Genotypes $(G)$ | 48.66 | 3 | 16.22 |
| Temperatures $(T)$ | 20.88 | 1 | 20.88 |
| Cultures $(C)$ | 2.46 | 2 | 1.23 |
| $G \times T$ | 11.48 | 3 | 3.83 |
| $G \times C$ | 1.70 | 6 | 0.28 |
| $T \times C$ | 1.14 | 2 | 0.57 |
| $G \times T \times C$ | 1.99 | 6 | 0.33 |
| Residual error | 9.67 | 96 | 0.10 |

4.

| | $\hat{V}_C(i)$ | $\hat{\beta}_i$ | $\hat{V}(I^*_{ij})$ |
|---|---|---|---|
| $P_1$ | 0.031 | −0.25 | 0.015 |
| $P_2$ | 0.608 | 1.59 | −0.014 |
| $F_1$ | 0.460 | 1.37 | 0.000 |
| $F_2$ | 0.328 | 1.17 | −0.001 |

| Source of variation | Sum of squares | Degrees of freedom | Mean square |
|---|---|---|---|
| Genotypes | 48.66 | 3 | 16.22 |
| Environments | 24.48 | 5 | 4.90 |
| Genotypes × environments | 15.17 | 15 | 1.01 |
| Regressions | 13.19 | 3 | 4.40 |
| Deviations from regressions | 1.98 | 12 | 0.16 |
| Residual error | 9.67 | 96 | 0.10 |

## Chapter 3

1. (a) $\frac{1}{2}$, (b) 11/32; they become (a) $\frac{1}{4}$, (b) 1/8.

2.

| $i$ | $b_i$ | $c_i$ | $d_i$ |
|---|---|---|---|
| 1 | 0.4 | −1.370 | −0.030 |
| 2 | 0.4 | 0.076 | 0.524 |
| 3 | −1.6 | 0.800 | 0.800 |
| 4 | 0.8 | 0.494 | −1.294 |

3. See Table 3.3.

4. $p_A = 0.373$, $p_B = 0.570$, $p_C = 0.057$.

| Phenotype | AA | BB | CC | AB | AC | BC |
|---|---|---|---|---|---|---|
| Expected | 122 | 286 | 3 | 374 | 37 | 57 |

5.

| Generation | 0 | 1 | 2 | 3 | 4 | 5 | 6 |
|---|---|---|---|---|---|---|---|
| Male frequency | 1 | 0 | 0.5 | 0.25 | 0.375 | 0.312 | 0.344 |
| Female frequency | 0 | 0 | 0 | 0.125 | 0.094 | 0.117 | 0.107 |

6.

| $t$ | 0 | 1 | 2 | 3 | 4 |
|---|---|---|---|---|---|
| | 0.5 | 0.5 | 0.375 | 0.312 | 0.281 |
| | 0.5 | 0.5 | 0.438 | 0.391 | 0.355 |
| | 0.5 | 0.5 | 0.250 | 0.250 | 0.250 |
| | 0.5 | 0.5 | 0.375 | 0.344 | 0.320 |

7. 0.0625; 0.00071875.

9.

| | | First locus | | |
|---|---|---|---|---|
| | | 0 | 1 | 2 | |
| Second locus | 0 | $(1+a)^2/16$ | $(1-a^2)/8$ | $(1-a)^2/16$ | $\frac{1}{4}$ |
| | 1 | $(1-a^2)/8$ | $(1+a^2)/4$ | $(1-a^2)/8$ | $\frac{1}{2}$ |
| | 2 | $(1-a)^2/16$ | $(1-a^2)/8$ | $(1+a)^2/16$ | $\frac{1}{4}$ |
| | | $\frac{1}{4}$ | $\frac{1}{2}$ | $\frac{1}{4}$ | |

## Chapter 4

1.

|  | BB | Bb | bb |
|---|---|---|---|
| A | $2q\,[a + d(q-p)]$ | $(q-p)[a + d(q-p)]$ | $-2p[a + d(q-p)]$ |
| D | $-2q^2 d$ | $2pqd$ | $-2p^2 d$ |

2.

|  | BB | Bb | bb |
|---|---|---|---|
| A | $2qa$ | $(q-p)\,a$ | $-2pa$ |
| D | $0$ | $d$ | $0$ |

3. They are all $\frac{1}{4}$.

4. $V_A = \frac{1}{64}$
   $V_D = \frac{1}{128}$
   $V_{AA} = \frac{1}{64}$
   $V_{AD} = \frac{1}{64}$
   $V_{DD} = \frac{1}{256}$

5. $a_1 = a_2 = d_1 = d_2 = \frac{1}{4}$; $aa_{12} = ad_{12} = ad_{21} = dd_{12} = -\frac{1}{4}$.

## Chapter 5

1. $\hat{m} = 0.294 \pm 0.014$
   $\hat{a} = 0.671 \pm 0.013$
   $\hat{d} = 0.001 \pm 0.028$

2. $\hat{m} = \phantom{0}92.8 \pm \phantom{0}4.8 \quad a\hat{a} = -19.8 \pm 4.6$
   $\hat{a} = \phantom{00}7.4 \pm \phantom{0}1.3 \quad a\hat{d} = \phantom{0}5.7 \pm 4.0$
   $\hat{d} = -28.2 \pm 12.2 \quad d\hat{d} = 21.5 \pm 7.9$

3. If the $a_i$s have mean value $\bar{a}$, then
$$a^2 = n^2 \bar{a}^2$$
$$\alpha = n\bar{a}^2 + \Sigma\,(a_i - \bar{a})^2.$$

Thus if the $a_i$s vary at all, $a^2/\alpha$ will underestimate the number of loci.

4. $n = 1$. No.

5. Let $y_i$ be the mean of the $i$th plot containing $F_2$ plants ($i = 1, \ldots, 10$) and compute
$$\bar{y} = \Sigma\,y_i/10$$
$$M = \Sigma\,(y_i - \bar{y})^2/9$$

$M$ has nine degrees of freedom and is distributed independently of the other variance components in Table 5.4 with Expected value
$$E(M) = \tfrac{1}{3}V_{1F_2} + V_{Eb} + \tfrac{1}{3}V_{Ew}.$$

The data can be re-analysed with this extra piece of information in the same way as before, but the results are unlikely to differ appreciably.

7. $V_e = \frac{1}{8}(\alpha + \delta) + V_E$.

## Chapter 6

1. No evidence of dominance or epistasis; $h^2 = \frac{1}{2}$. Environmental variances are estimated as 4.6 and 6.2 in males and females, in good agreement with values within inbred lines. The environmental deviation can be partitioned into a local component acting independently in the two sternites, and a component common to both sternites, with the following variances:

|                  | Males | Females |
|------------------|-------|---------|
| Local component  | 1.2   | 1.2     |
| Common component | 3.4   | 5.0     |

(Compare Problem 5),

2. $\hat{\rho}_H = 0.065 \pm 0.029$, $\hat{\rho}_F = 0.189 \pm 0.034$, $\text{Cov}(\hat{\rho}_H, \hat{\rho}_F) = 0.000582$. 95 per cent confidence limits for $h^2$ based on $\hat{\rho}_H$ are 0.03 to 0.49. $(\hat{\rho}_F - 2\hat{\rho}_H) = 0.059 \pm 0.058$, not significant. Assuming that $\rho_F = 2\rho_H$, $h^2 = 0.378 \pm 0.068$, 95 per cent limits are 0.24 to 0.51.

3. Write $d_i = y_{1i} - y_{2i}$, $s_i = y_{1i} + y_{2i}$.

| Source of variation | Sum of squares | Degrees of freedom | Expected mean square |
|---------------------|----------------|---------------------|----------------------|
| Between pairs | $\frac{1}{2}(s_i - \bar{s})^2$ | $N - 1$ | $(1 + \rho)V$ |
| Without pairs | $\frac{1}{2}\sum d_i^2$ | $N$ | $(1 - \rho)V$ |

If $M_B$ and $M_W$ are the mean squares between and within pairs, estimate $\rho$ as $(M_B - M_W)/(M_B + M_W)$.

4.

|                                        | Source of variation | Mean square | Degrees of freedom |
|----------------------------------------|---------------------|-------------|---------------------|
| Monozygotic twins brought up apart     | Between pairs       | 29.1        | 39                  |
|                                        | Within pairs        | 7.1         | 40                  |
| Monozygotic twins brought up together  | Between pairs       | 18.3        | 42                  |
|                                        | Within pairs        | 8.0         | 43                  |
| Dizygotic twins                        | Between pairs       | 19.1        | 24                  |
|                                        | Within pairs        | 14.1        | 25                  |

Estimates of $V$ and $\rho$ are as follows:

|                                       | $\hat{V}$       | $\hat{\rho}$     |
|---------------------------------------|-----------------|------------------|
| Monozygotic twins brought up apart    | $18.1 \pm 3.3$  | $0.61 \pm 0.10$  |
| Monozygotic twins brought up together | $13.2 \pm 2.1$  | $0.39 \pm 0.13$  |
| Dizygotic twins                       | $16.6 \pm 3.3$  | $0.15 \pm 0.19$  |

No evidence of any difference in phenotypic variance; no evidence that correlation between monozygotic twins is due to common family environment. We may estimate the total genetic component of the character as contributing about 50 per cent of the variance, but the evidence from the small number of dizygotic twins is too inaccurate to allow further conclusions about the mode of inheritance.

5. $\hat{V}_X = 208$, $(\hat{V}_G + \hat{V}_P) = 60$, $\hat{V}_T = 22$, $\hat{r} = 0.73$.

$V_X$ represents differences between years and between ages of rams, which are confounded in this experiment.

6. (a) $V_A = 15$, $V_{AA} = 58$, $V_E = 27$; (b) $V_A = 44$, $V_{AAAAA..} = 29$, $V_E = 27$.

|  | (a) | (b) |
|---|---|---|
| Half sibs | 0.074 | 0.110 |
| Cousins | 0.028 | 0.055 |

7. If we assume no dominance, the weighted least squares estimates of the genetic and environmental components of variance are:

|  | Variance (plant height) | Covariance | Variance (ear height) | Correlation |
|---|---|---|---|---|
| Genetic | 34.2 | 17.2 | 14.4 | $0.78 \pm 0.04$ |
| Environmental | 8.3 | 3.2 | 3.3 | $0.62 \pm 0.05$ |

(NB Mode and Robinson (1959) give the standard error of the additive genetic correlation in Table 6.10b as 0.22, not 0.14, but I have been unable to reproduce their value.)

8. 0.125; 0.275; 0.275; 0.227.

# Chapter 7

1. Inbreeding coefficients: 0, 0; 0.375, 0.5; 0.5, 0.594; 0.375, 0; 0.5, 0; 0, 0. Regression: Litter size $= 8.3 - 2.1 f$(mother) $- 2.7 f$(young).

2. $\delta_{1i} = -2p_i q_i d_i$. $\delta_1 = \sum \delta_{1i}$ = regression of mean value on $f$, $V_D = \sum \delta_{1i}^2$; $\delta_1^2 / V_D$ can be used to estimate the number of loci if the $\delta_{1i}$s are of the same sign and magnitude.

4.

| Source of variation | Sum of squares | Degrees of freedom | Mean square | Expected mean square |
|---|---|---|---|---|
| General combining abilities | 1368 | 8 | 171.0 | $14V_g + V_e$ |
| Specific combining abilities | 288 | 27 | 10.7 | $2V_s + V_e$ |
| Reciprocal effects | 145 | 36 | 4.0 | $2V_r + V_e$ |

If we assume that $V_r = 0$, $\hat{V}_g = 11.9$, $\hat{V}_s = 3.8$.
With reciprocal effects averaged the sums of squares will be halved; $V_e$ is also halved since the unit of measurement has changed, and can be estimated as 2.0.

5. $E\sum \hat{g}_i^2 = \sum g_i^2 + \dfrac{(p-1)^2}{p(p-2)} V_e$

$E\sum \hat{m}_i^2 = \sum m_i^2 + \dfrac{2(p-1)}{p} V_e$

$E \sum_{i<j} \hat{s}_{ij}^2 = \sum_{i<j} s_{ij}^2 + \tfrac{1}{4}p(p-3)V_e.$

6. $\hat{V}_a = 90.2$
$\hat{V}_d = 4.5$
$\hat{V}_s = 3.3$.

## Chapter 8

1. $Y_C = \dfrac{2(1 - h^2)^2}{(2 - h^4)} + \dfrac{h^2(2 - h^2)}{(2 - h^4)} Y_P + \dfrac{2h^2(1 - h^2)}{(2 - h^4)} Y_G + e$

$\text{Var}(e) = \frac{1}{2}h^4(3 - 2h^2)/(2 - h^4)$.

2. 0.0412.

3. (a) $V_Y = 86$, $h^2 = 0.53$; (b) $V_Y = 89$, $h^2 = 0.55$.

5. $E(G_C | G_P) = G_P$.
$\text{Var}(G_C | G_P) \simeq (nG_P - \frac{1}{2}G_P^2 - \frac{1}{2}npq)/(2n - 1)$.

6. $E(G_C | G_P) \simeq \frac{1}{2}pq^2/(1 + q) + 2G_P/(1 + q) - G_P^2/n(1 + q)^2$.

7.

| $G_F$ | 0 | $d$ | $2d$ | 1 | $1 + d$ | 2 |
|-------|------|-------------------------------|---------|-------|-------------------------------|----------------|
| $E(G_C | G_F)$ | $2pd$ | $(p + \frac{1}{2})d + \frac{1}{2}p$ | $d + p$ | $d + p$ | $(q + \frac{1}{2})d + 1\frac{1}{2}p$ | $2qd + 2p$ |

8. $E(Y_C | Y_F) = 0.0442 + 0.2812\, Y_F - 0.0442\, Y_F^2$.

## Chapter 9

1. (a) 1.755, (b) 1.648, (c) 1.638.

2. 7.0.

3. $m(t + 1) = m(t) + \dfrac{\frac{1}{2}[\theta - m(t)]V_A(t)}{V_A(t) + \gamma + V_D + V_E}$

$V_A(t + 1) = \frac{1}{2}V_A(t) - \frac{1}{4}\dfrac{V_A^2(t)}{V_A(t) + \gamma + V_D + V_E} + \frac{1}{2}V_A$.

4. (a) 78, (b) 67.

5. $\beta = 0.0214$. (a) 71, (b) 41.

6. $R_{AA}^* = \frac{1}{2}(1 - \bar{r})S_{AA}$, where $\bar{r}$ is the average recombination fraction. (It is assumed that $r_{ij}$ is unrelated to the size of the interaction between loci $i$ and $j$.)

## Chapter 10

2. Type of selection          B          C

| | B | C |
|---|---|---|
| Stabilizing rank | $-0.9785/V$ | $0.9785/V$ |
| Stabilizing value | $-0.9785/V$ | 0 |
| Disruptive rank | $2.25/V$ | $-2.25/V$ |
| Disruptive value | $2.25/V$ | 0 |

where $V$ is the phenotypic variance.

$V_G = 5.25$, $\hat{V}_G = 3.59$, $\bar{e}' = \bar{e}$, $d_i' = 1.013 d_i$. Gene frequencies at $\frac{1}{2}$ should stay there, frequencies at $\frac{1}{4}$ should move towards 0 at a rate of 0.003 per generation, frequencies at $\frac{3}{4}$ should move towards 1 at the same rate.

3. $V_G = 0.33 - V_E$.

4. (a) 0.375, (b) 0.012.

5. $h^2 = 0.23$.

## Chapter 11

1. $\hat{W} = \text{constant} + 0.45 Y_1 + 0.31 Y_2 + 1.01 Y_3$
   $\rho(W, \hat{W}) = 0.89$.

2. $\hat{W} = \text{constant} + Y_2 + 0.48 Y_3^*$
   $\rho(W, \hat{W}) = 0.22$.

3. The predicted breeding values (in standard deviations above the mean) are 0.4 and 0.45 respectively.

4. $\hat{A} = m + 0.467(Y_1 - m) + 0.133(Y_2 - m)$.

5. $\hat{A} = m + b_1(Y_1 - m) + b_2(Y_2 - m)$
   $b_1 = h^2(1 - \frac{1}{2}h^2 + \frac{1}{4}nh^2)/(1 - \frac{1}{2}h^2)(1 + \frac{1}{2}nh^2)$
   $b_2 = \frac{1}{2}nh^2(1 - h^2)/(1 - \frac{1}{2}h^2)(1 + \frac{1}{2}nh^2)$.

6. In the same notation,

$$
\begin{bmatrix} b_1 \\ b_2 \end{bmatrix} = \begin{bmatrix} [1 + \frac{1}{2}h^2 + \frac{1}{4}h^2(m-1)]/m & \frac{1}{4}h^2 \\ \frac{1}{4}h^2 & [1 + \frac{1}{4}h^2(n-m-1)]/(n-m) \end{bmatrix}^{-1} \begin{bmatrix} \frac{1}{2}h^2 \\ \frac{1}{2}h^2 \end{bmatrix}
$$

7. 

| Number of varieties tested | 100 | 50 | 25 | 20 | 10 |
|---|---|---|---|---|---|
| Expected genetic gain | 1.12 | 1.30 | 1.40 | 1.39 | 1.30 |

Test 25 varieties with four replications or 20 varieties with five replications.

8. Test six daughters from each of 33 bulls.

## Chapter 12

1. $N_e = 13.3$. Mean litter size $= 6.8$.

2. Var('true' mean of line) $= 31.8$, Var(environmental contribution to mean) $= 3.3$, Var(difference between two observed means) $= 70.2$; Prob($|$difference$|$exceeds 10) $= 0.23$.

3.

| $N$ | 20 | 100 | 500 |
|---|---|---|---|
| $p$ | | | |
| 0.1 | 0.109 | 0.151 | 0.396 |
| 0.5 | 0.525 | 0.622 | 0.924 |
| 0.9 | 0.909 | 0.939 | 0.996 |

4. 240; 28 generations.

# References

Anderson, T. W. (1958). *An introduction to multivariate statistical analysis*. Wiley, New York.

Avery, P. J. and Hill, W. G. (1977). Variability in genetic parameters among small populations. *Genet. Res.* **29**, 193–213.

Bailey, N. T. J. (1961). *Introduction to the mathematical theory of genetic linkage*. Clarendon Press, Oxford.

—— (1964). *The elements of stochastic processes with applications to the natural sciences*. Wiley, New York.

Bateson, W., Saunders, E. R., and Punnett, R. C. (1905). Further experiments on inheritance in sweet peas and stocks: preliminary account. *Proc. R. Soc.* **B77**, 236–8.

Baur, E. (1907). Untersuchungen über die Erblichkeitsverhältnisse einer nur in Bastardform lebenshäfigen Sippe von *Antirrhinum majus*. *Ber. dt. bot. Ges.* **25**, 442–54.

*Biometrika tables for statisticians* (eds. E. S. Pearson and H. O. Hartley) Vol. 1 (1970, 3rd edn), Vol. 2 (1972). Cambridge University Press.

Box, J. F. (1978). *R. A. Fisher: the life of a scientist*. Wiley, New York.

Boyle, C. R. and Elston, R. C. (1979). Multifactorial genetic models for quantitative traits in humans. *Biometrics* **35**, 55–68.

Bregger, T. (1918). Linkage in maize: the C aleurone factor and waxy endosperm. *Am. Nat.* **52**, 57–61.

Bulmer, M. G. (1970). *The biology of twinning in man*. Clarendon Press, Oxford.

—— (1971). Stable equilibria under the two-island model. *Heredity, Lond.* **27**, 321–30.

—— (1974a). Linkage disequilibrium and genetic variability. *Genet. Res.* **23**, 281–9.

—— (1974b). Density-dependent selection and character displacement. *Am. Nat.* **108**, 45–48.

—— (1976a). Regressions between relatives. *Genet. Res.* **28**, 199–203.

—— (1976b). The effect of selection on genetic variability: a simulation study. *Genet. Res.* **28**, 101–17.

Carter, C. O. (1965). The inheritance of common congenital malformations. *Prog. med. Genet.* **4**, 59–84.

Castle, W. E. (1903). The laws of heredity of Galton and Mendel, and some laws governing race improvement by selection. *Proc. Am. Acad. Arts Sci.* **39**, 223–42.

Clayton, G. A., Morris, J. A., and Robertson, A. (1957). An experimental check on quantitative genetic theory. I. Short-term responses to selection. *J. Genet.* **55**, 131–51.

Cochran, W. G. (1951). Improvement by means of selection. In *Proc. 2nd Berkeley Symp. Math. Stat. Prob.* (ed. J. Neyman) pp. 449–70. University of California Press, Berkeley.

Comstock, R. E. and Robinson, H. F. (1952). Estimation of average dominance of genes. In *Heterosis* (ed. J. W. Gowen) pp. 494–516. Iowa State College Press, Ames.

Cotterman, C. W. (1940). A calculus for statistico-genetics. In *Genetics and social structure*, Benchmark Papers in Genetics, Vol. 1 (ed. P. A. Ballonoff) pp. 157–272. 1974. Dowden, Hutchinson, and Ross, Stroudsburg, Pennsylvania.

Crow, J. F. and Kimura, M. (1970). *An introduction to population genetics theory.* Harper and Row, New York.

Cunningham, E. P. (1975). Multi-stage index selection. *Theor. appl. Genet.* **46**, 55–61.

Curnow, R. N. (1961a). The estimation of repeatability and heritability from records subject to culling. *Biometrics* **17**, 553–66.

—— (1961b). Optimal programmes for varietal selection. *J. R. statist. Soc.* **B23**, 282–318.

—— and Smith, C. (1975). Multifactorial models for familial diseases in man. *J. R. statist. Soc.* **A138**, 131–69.

Draper, N. R. and Smith, H. (1966). *Applied regression analysis.* Wiley, New York.

Dunlop, A. A. (1962). Interactions between heredity and environment in the Australian merino. I. Strain × location interactions in wool traits. *Aust. J. agric. Res.* **13**, 503–31.

Eanes, W. R. (1978). Morphological variance and enzyme heterozygosity in the monarch butterfly. *Nature, Lond.* **276**, 263–4.

East, E. M. (1916). Studies on size inheritance in *Nicotiana*. *Genetics, Princeton* **1**, 164–76.

Eberhart, S. A. and Gardner, C. O. (1966). A general model for genetic effects. *Biometrics* **22**, 864–81.

—— and Russell, W. A. (1966). Stability parameters for comparing varieties. *Crop. Sci.* **6**, 36–40.

Eisen, E. J., Bohren, B. B., and McKean, H. E. (1966). Sex-linked and maternal effects in the diallel cross. *Aust. J. biol. Sci.* **19**, 1061–71.

Fairfield Smith, H. (1936). A discriminant function for plant selection. *Ann. Eugen.* **7**, 240–50.

Falconer, D. S. (1952). The problem of environment and selection. *Am. Nat.* **86**, 293–8.

—— (1960). *Introduction to quantitative genetics.* Oliver and Boyd, Edinburgh.

—— (1977). Some results of the Edinburgh selection experiments with mice. *Proc. Int Conf. Quant Genet.* (eds. E. Pollak, O. Kempthorne, and T. B. Bailey) pp. 101–15. Iowa State University Press, Ames.

Felsenstein, J. (1977). Multivariate normal genetic models with a finite number of loci. *Proc. Int. Conf. Quant. Genet.* (eds. E. Pollak, O. Kempthorne, and T. B. Bailey) pp. 227–46. Iowa State University Press, Ames.

Finlay, K. W. and Wilkinson, G. N. (1963). The analysis of adaptation in a plant-breeding programme. *Aust. J. agric. Res.* **14**, 742–54.

Finney, D. J. (1958). Statistical problems of plant selection. *Bull. Int. Statist. Inst.* **36**, 242–68.

Fisher, R. A. (1918). The correlation between relatives on the supposition of Mendelian inheritance. *Trans. R. Soc. Edin.* **52**, 399–433.

—— (1921). On the 'probable error' of a coefficient of correlation deduced from a small sample. *Metron, Rovigo* **1**, part 4, 1–32.

—— (1930). *The genetical theory of natural selection.* Clarendon Press, Oxford.

—— (1946). *Statistical methods for research workers*, 10th edn. Oliver and Boyd, Edinburgh.

—— (1952). Statistical methods in genetics. *Heredity, Lond.* **6**, 1–12.

—— and Yates, F. (1953). *Statistical tables for biological, agricultural and medical research*, 4th edn. Oliver and Boyd, London.

Fleming, W. H. (1979). Equilibrium distributions of continuous polygenic traits. *SIAM J. appl. Math.* **36**, 148–68.

Frankham, R. (1977a). The nature of quantitative genetic variation in *Drosophila*. III. Mechanism of dosage compensation for sex-linked abdominal bristle polygenes. *Genetics, Princeton* **85**, 185–91.

—— (1977b). Optimum selection intensities in artificial selection programmes: an experimental evaluation. *Genet. Res.* **30**, 115–19.

Fredeen, H. T. and Jonsson, P. (1957). Genic variance and covariance in Danish Landrace swine as evaluated under a system of individual feeding of progeny test groups. *Z. Tierzücht. ZüchtBiol.* **70**, 348–63.

Galton, F. (1887). *Natural inheritance*. Macmillan, London.

Garrod, A. E. (1909). *Inborn errors of metabolism*. Oxford University Press, London.

Graybill, F. A., Martin, E., and Godfrey, G. (1956). Confidence intervals for variance components specifying genetic heritability. *Biometrics* **12**, 99–109.

Griffing, B. (1956). Concept of general and specific combining ability in relation to diallel crossing systems. *Aust. J. biol. Sci.* **9**, 463–93.

—— (1960). Theoretical consequences of truncation selection based on the individual phenotype. *Aust. J. biol. Sci.* **13**, 307–43.

—— and Langridge, J. (1963). Phenotypic stability of growth in the self-fertilized species, *Arabidopsis thaliana*. In *Statistical genetics and plant breeding* (eds. W. D. Hanson and H. F. Robinson) pp. 368–94. National Academy of Sciences, Washington, DC.

Haldane, J. B. S. (1954). The measurement of natural selection. *Proc. IX Int. Congr. Genet., Caryologia* (Suppl.) pp. 480–7.

Hardy, G. H. (1908). Mendelian proportions in a mixed population. *Science, N.Y.* **28**, 49–50.

Harris, D. L. (1963). The influence of errors of parameter estimation upon index selection. In *Statistical genetics and plant breeding* (eds. W. D. Hanson and H. F. Robinson) pp. 491–500. National Academy of Sciences, Washington, DC.

—— (1964). Expected and predicted progress from index selection involving estimates of population parameters. *Biometrics* **20**, 46–72.

Harris, H. (1975). *The principles of human biochemical genetics*. North-Holland, Amsterdam.

Harville, D. A. (1977). Maximum likelihood approaches to variance component estimation and to related problems. *J. Am. Stat. Ass.* **72**, 320–38.

Hayman, B. I. (1954). The theory and analysis of diallel crosses. *Genetics, Princeton* **39**, 789–809.

—— (1960). Maximum likelihood estimation of genetic components of variation. *Biometrics* **16**, 369–81.

Hazel, L. N. (1943). The genetic basis for constructing selection indices. *Genetics, Princeton* **38**, 476–90.

—— and Lush, J. L. (1942). The efficiency of three methods of selection. *J. Hered.* **33**, 393–9.

Henderson, C. R. (1963). Selection index and expected genetic advance. In *Statistical genetics and plant breeding* (eds. W. D. Hanson and H. F. Robinson) pp. 141–63. National Academy of Sciences, Washington, DC.

—— (1973). Sire evaluation and genetic trends. In *Proc. Anim. Breed. Genet. Symp., Blacksburg, Virginia*, pp. 10–41. American Society of Animal Sciences, Champaign, Illinois.

—— (1977). Prediction of future records. In *Proc. Int. Conf. Quant. Genet.* (eds. E. Pollak, O. Kempthorne, and T. B. Bailey) pp. 615–38. Iowa State University Press, Ames.

Hill, W. G. (1977). Variation in response to selection. *Proc. Int. Conf. Quant. Genet.* (eds. E. Pollak, O. Kempthorne, and T. B. Bailey) pp. 343–65. Iowa State University Press, Ames.

Hinkelmann, K. (1977). Diallel and multi-cross designs: What do they achieve? *Proc. Int Conf. Quant. Genet.* (eds. E. Pollak, O. Kempthorne, and T. B. Bailey) pp. 659–76. Iowa State University Press, Ames.

Holt, S. B. (1968). *The genetics of dermal ridges.* Thomas, Springfield, Ill.

Jacquard, A. (1974). *The genetic structure of populations* (trs. D. and B. Charlesworth). Springer-Verlag, Berlin.

Johannsen, W. (1903). *Über Erblichkeit in Populationen und in reinen Linien.* Gustav Fischer, Jena.

—— (1909). *Elemente der exakten Erblichkeitslehre.* Gustav Fischer, Jena.

Johansson, I. and Korkman, N. (1951). A study of the variation in production traits of bacon pigs. *Acta agric. scand.* **1**, 62–96.

—— and Rendel, J. (1968). *Genetics and animal breeding.* Oliver and Boyd, Edinburgh.

Jones, G. L., Matzinger, D. F., and Collins, W. K. (1960). A comparison of flux-cured tobacco varieties repeated over locations and years with implications on optimum plot allocation. *Agron. J.* **52**, 195–9.

Kallmann, F. J. and Reisner, D. (1943). Twin studies on the significance of genetic factors in tuberculosis. *Am. Rev. Tuberc. pulm. Dis.* **47**, 549–74.

Kemeny, J. G. and Snell, J. L. (1960). *Finite Markov chains.* Van Nostrand, Princeton.

Kempthorne, O. and Nordskog, A. W. (1959). Restricted selection indices. *Biometrics* **15**, 10–19.

Kendall, M. G. and Stuart, A. (1963). *The advanced theory of statistics*, 2nd edn, Vol. 1. Griffin, London.

Kimura, M. (1965). A stochastic model concerning the maintenance of genetic variability in quantitative characters. *Proc. natn. Acad. Sci. U.S.A.* **54**, 731–6.

Lande, R. (1976). The maintenance of genetic variability by mutation in a polygenic character with linked loci. *Genet. Res.* **26**, 221–35.

—— (1977). The influence of the mating system on the maintenance of genetic variability in polygenic characters. *Genetics, Princeton* **86**, 485–98.

Lerner, I. M. (1954). *Genetic homeostasis.* Oliver and Boyd, Edinburgh.

—— (1958). *The genetic basis of selection.* Wiley, New York.

Li, C. C. (1955). *Population genetics.* Chicago University Press.

Lindley, D. V. (1947). Regression lines and the linear functional relationship. *J. R. statist. Soc.* Suppl. **9**, 218–44.

Malécot, G. (1948). *Les mathématiques de l'hérédité.* Masson, Paris.

Mather, K. and Jinks, J. L. (1971). *Biometrical genetics.* Chapman and Hall, London.

—— (1977). *Introduction to biometrical genetics.* Chapman and Hall, London.

Mayo, O. (1980). *An introduction to the theory of plant breeding.* Clarendon Press, Oxford.

Mendel, G. (1866). Versuche über Pflanzenhybriden. *Verh. Naturf. Ver. Brünn* **4**, 3–44. [For an English translation see Stern and Sherwood 1966.]

Miller, P. A., Williams, J. C., and Robinson, H. F. (1959). Variety × environment interactions in cotton variety tests and their implications on testing methods. *Agron. J.* **51**, 132–4.

Mode, C. J. and Robinson, H. F. (1959). Pleiotropism and the genetic variance and covariance. *Biometrics* **15**, 518–37.

Moll, R. H., Lindsey, M. F., and Robinson, H. F. (1964). Estimates of genetic variances and level of dominance in maize. *Genetics, Princeton* **49**, 411–23.

National Bureau of Standards (1959). Tables of the bivariate normal distribution function and related functions. (Applied Mathematics Series, 50). US Government Printing Office, Washington.

Neal, N. P. (1935). The decrease in yielding capacity in advanced generations of hybrid corn. *J. Am. Soc. Agron.* **27**, 666–70.

Owen, D. B. (1962). *Handbook of statistical tables*. Addison-Wesley, Reading, Mass.

Pearson, K. (1904). Mathematical contributions to the theory of evolution. XII. On a generalized theory of alternative inheritance, with special reference to Mendel's laws. *Phil. Trans. R. Soc.* **A203**, 53–86.

—— and Lee, A. (1903). Laws of inheritance in man. *Biometrika* **2**, 357–62.

Pederson, D. G. (1969). The prediction of selection response in a self-fertilising species. *Aust. J. biol. Sci.* **22**, 117–29, 1245–57.

Pirchner, F. (1969). *Population genetics in animal breeding*. Freeman, San Francisco.

Powers, L. (1951). Gene analysis by the partitioning method when interactions of genes are involved. *Bot. Gaz.* **113**, 1–23.

Punnett, R. C. (1917). Reduplication series in sweet peas. II. *J. Genet.* **6**, 185–93.

Rao, C. R. (1965). *Linear statistical inference and its applications*. Wiley, New York.

Rasmusson, D. C. and Lambert, J. W. (1961). Variety × environment interactions in barley variety tests. *Crop Sci.* **1**, 261–2.

Robertson, A. (1957). Optimum group size in progeny testing and family selection. *Biometrics* **13**, 442–50.

—— (1960). A theory of limits in artificial selection. *Proc. R. Soc.* **B153**, 234–49.

—— (1966). A mathematical model of the culling process in dairy cattle. *Anim. Prod.* **8**, 95–108.

—— (1970). A theory of limits in artificial selection with many linked loci. In *Mathematical topics in population genetics* (ed. K. Kojima) pp. 246–88. Springer-Verlag, Berlin.

—— (1977). The non-linearity of offspring–parent regression. In *Proc. Int. Conf. Quant. Genet.* (eds. E. Pollack, O. Kempthorne, and T. B. Bailey) pp. 297–304. Iowa State University Press, Ames.

Sales, J. and Hill, W. G. (1976). Effect of sampling errors on efficiency of selection indices. *Anim. Prod.* **22**, 1–17; **23**, 1–14.

Scheffé, H. (1959). *The analysis of variance*. Wiley, New York.

Searle, S. R. (1971). *Linear models*. Wiley, New York.

Shields, J. (1962). *Monozygotic twins brought up apart and brought up together*. Oxford University Press, London.

Slatkin, M. (1978). Spatial patterns in the distribution of polygenic characters. *J. theor. Biol.* **70**, 213–28.

Snedecor, G. W. and Cochran, W. G. (1969). *Statistical methods*, 6th edn. Iowa State University Press, Ames.

Sprague, G. F. and Tatum, L. A. (1942). General versus specific combining ability in single crosses of corn. *J. Am. Soc. Agron.* **34**, 923–32.

Stern, C. and Sherwood, E. R. (1966). *The origin of genetics*. Freeman, San Francisco.

Strickberger, M. W. (1976). *Genetics*, 2nd edn. Macmillan, New York.

Turner, H. N. and Young, S. S. Y. (1969). *Quantitative genetics in sheep breeding*. Macmillan, Melbourne.

Watson, J. D. and Crick, F. H. C. (1953). A structure for deoxyribose nucleic acid. *Nature, Lond.* **171**, 737–8.

Weinberg, W. (1908). Uber den Nachweis der Vererbung beim Menschen. *Jh. Ver. vaterl. Naturk. Württ.* **64**, 369–82.

Whitehouse, H. L. K. (1973). *Towards an understanding of the mechanism of heredity*, 3rd edn. Arnold, London.

Wright, S. (1931). Evolution in Mendelian populations. *Genetics, Princeton*, **16**, 97–159.

Young, S. S. Y. (1961). A further examination of the relative efficiency of three methods of selection under less-restricted conditions. *Genet. Res.* **2**, 106–21.

Yule, G. U. (1902). Mendel's laws and their probable relations to intra-racial heredity. *New Phytol.* **1**, 193–207, 222–38.

—— (1906). On the theory of inheritance of quantitative compound characters on the basis of Mendel's laws—a preliminary note. *Report 3rd Int. Conf. Genetics*, pp. 140–2.

# Author index

# Subject Index

DATE DUE